M & E TECBOOKS

The primary need of students taking Technician Education Council (TEC) courses is for books which will reflect the new methodology and syllabus requirements. In presenting M & E TECBOOKS, we believe that the careful collaboration between subject editors and authors has resulted in the very best, "tailor-made" texts which could be devised for the student—and which are likely to be of guidance to the lecturer also. The aim has been to provide books based on TEC's own objectives, concisely and authoritatively presented, priced as closely as possible to the student budget.

The M & E TECBOOK Series

Electrical and Electronic Principles Level II

FRANK GOODALL

Ph.D., M.Sc., C.Eng., M.I.E.R.E.

Vice Principal, Granville College, Sheffield

and

D. K. RISHTON

M.Sc., C.Eng., M.I.E.E., A.R.T.C.S.

Principal Lecturer in Electrical Engineering, Salford College of Technology

Macdonald & Evans Ltd.
Estover, Plymouth PL6 7PZ

First published 1983

British Library Cataloguing in Publication Data

Goodall, F.
Electrical and electronic principles
Level II.—(The M & E TECBOOK series)
1. Electronic apparatus and appliances
I. Title II. Rishton, D. K.
621.381 TK7870

ISBN 0-7121-0592-1

Text set in 10/11 pt Linotron 202 Times,
printed and bound in
Great Britain at
The Pitman Press, Bath

Preface

Conceived after the Haselgrove report in 1969 and established in 1973, TEC, the Technician Educational Council, undertook the most searching reappraisal of courses provided in the F.E. sector for Engineering Technicians. As part of this reappraisal, syllabus content was completely rethought, and standard units were produced by TEC for use in the modular structure courses.

Previously, a syllabus indicated the area of study, but left the depth of treatment to the professional judgment of the lecturer involved. TEC units on the other hand indicate quite specific topics and depth of treatment.

This book has therefore been written to cover exactly the standard unit of Electrical and Electronic Principles II not necessarily in sequence but nevertheless fully.

Wherever possible the simplest explanation has been given and analogies taken from very simple everyday experiences have been used. Also, where some mathematical principles have had to be included, those principles have been explained as part of the chapter without need to turn to a second textbook.

Many worked examples have been included, but once again only to the depth of treatment which is necessary. Each chapter is terminated by a series of examples, which should be attempted by the reader to help consolidate his understanding. In particular, different types of question are included, to help develop a knowledge of different types, all of which are likely to be part of TEC assessments in colleges.

In addition, part way through and at the end of the text, two more comprehensive progress text papers are given which, where possible, should be completed carefully in full.

Thanks are due to all who have made suggestions or have made useful contributions in discussions.

Acknowledgment is also given to the Post Office Educational pamphlets from which ideas for several of the diagrams were developed.

1983

F.G.
D.K.R.

Contents

Preface v

List of Illustrations ix

1 Units and Circuits 1

Introduction; Basic electrical units; Mechanical units; Multiples and sub-multiples of units; Circuits; The superposition theorem; Kirchoff's laws; Self-assessment questions

2 Magnatic Fields, Circuits and Effects 24

Revision of basic principles; Simple electromagnetic fields; The relationship between simple magnetic units; Magnetising curves for typical ferromagnetic materials; Self-assessment questions

3 Capacitors and Capacitance 43

Electric fields and electric charges; The capacitor; Series and parallel combinations of capacitors; Types of capacitor; Capacitors in simple circuits; Self-assessment questions

4 Electromagnetic Induction 68

Faraday's discoveries; The generator principle; The motor principle; Self inductance; Arcing in inductive circuits; The inductance of a coil in relation to its physical properties; Mutual inductance; The basic principle of the transformer; Self-assessment questions; Cumulative questions, chapters 1–4

5 Alternating Voltages and Currents 94

Mathematical considerations; Alternating and unidirectional waveforms; Instantaneous, average and r.m.s. values; Phasor and algebraic representation of alternating quantities; The principle of rectification; Half-wave and full-wave rectification; The relationships between graphical, phasor and algebraic representation of sine waves; Self-assessment questions

6 Single Phase a.c. Circuits 118

Introduction; A pure resistance in an a.c. single phase circuit; A pure inductance in a simple a.c. circuit; A pure capacitance in a simple a.c. circuit; Power dissipated in a purely inductive or capacitive circuit; Dependence of inductive reactance on frequency; Dependence of capacitive reactance on frequency; Series combinations of components in a.c. circuits; The power dissipated in L–R and L–C circuits; The series resonant condition; Calculation of power in a.c. circuits for sinusoidal waveforms; Self-assessment questions

7 Measurement and Measuring Devices 140

The general requirements for instruments; The correct connection of ammeters and voltmeters; Measurement errors; Extension of meter ranges; The cathode ray oscilloscope as a measuring device; The limitations of moving-coil and moving-iron galvanometers; Null measurement techniques; The Wheatstone bridge; The d.c. potentiometer; The ohm meter; Digital meters; Self-assessment questions

8 Semiconductor Diodes and Transistors 160

Atomic structure; Semiconductors; The effect of temperature on conductors, insulators and semiconductors; The structure of germanium and silicon; Intrinsic semiconduction; Doped semiconductors; The pn junction; The bipolar transistor; The basic transistor action; Transistor configurations; Self-assessment questions; Cumulative questions, chapters 5–8

Appendix: Answers to Self-assessment and Cumulative Questions 190

Index 192

List of Illustrations

1. Potential difference across a load. 2
2. A simple illustration of torque. 5
3. Voltage and current distribution in a series circuit. 7
4. Voltage and current distribution in a parallel circuit. 8
5. Circuit diagram for Ohm's law solution. 9
6. The basic principle of superposition. 13
7. Circuit diagram for superposition solution. 14
8. The principle of Kirchoff's first law. 18
9. Circuit diagram for solution of problem involving Kirchoff's law. 18
10. Final solution of problem involving Kirchoff's law. 22
11. Magnetic field around a single conductor. 25
12. Comparison of the magnetic fields around: (*a*) a bar magnet; and (*b*) a solenoid. 26
13. A simple coil electromagnet, showing its important dimensions. 28
14. Comparison of electromagnetic and electrical terms. 31
15. Changes in flux density $\boldsymbol{B}$ resulting from changes in magnetising force $\boldsymbol{H}$. 32
16. Typical curve showing variation in permeability with increasing $\boldsymbol{H}$. 34
17. Typical maximum μ_r values for common magnetic materials. 34
18. Three examples of magnetic screening: (*a*) and (*b*) low frequency currents; (*c*) high frequency currents. 38
19. Cyclical magnetisation showing hysteresis effect. 39
20. Hysteresis loops for: (*a*) a soft magnetic material and (*b*) a hard magnetic material. 40
21. Hysteresis loop for a ferrite. 41
22. The electrostatic field between charges. 44
23. The action of charging a capacitor. 45
24. The electrostatic field between capacitor plates. 46
25. Charging current and capacitor voltage curves. 46
26. The principle of energy storage in a capacitor dielectric showing electron orbits with: (*a*) plates charged; and (*b*) plates uncharged. 49
27. The air dielectric variable capacitor. 53
28. Voltage distribution for capacitors in series. 54
29. Current distribution for capacitors in parallel. 55
30. Calculation of the total value of capacitance. 56

31. Relationship between p.d. and charge. 58
32. Construction of the wax paper/foil type of capacitor. 60
33. An example of the mica type of capacitor. 60
34. Examples of the ceramic type of capacitor: (*a*) disc type; (*b*) hat (or cup) type; and (*c*) tube type. 61
35. Examples of the: (*a*) wet; and (*b*) dry types of capacitor. 62
36. Circuit diagram for Example. 64
37. Conductor movement in a magnetic field. 70
38. Conductors of different lengths cutting a magnetic field. 71
39. Fleming's right hand rule. 72
40. E.m.f. induced in a rotating coil. 73
41. Coil rotating at constant speed in a magnetic field. 74
42. Curve produced by plotting the variation in e.m.f. induced in the coil in Fig. 41. 75
43. Force exerted on a current-carrying conductor in a magnetic field. 76
44. Fleming's left hand rule. 78
45. Growth of current in an inductive circuit. 80
46. Induced e.m.f. caused by mutual inductance effect. 86
47. The definition of the radian. 95
48. Ilustration of the use of the mid-value as an approximation. 95
49. The mid-ordinate rule for area calculation. 96
50. Basic trigonometrical terms. 97
51. (*a*) Constant d.c. current; (*b*) d.c. current halved. 98
52. (*a*) Alternating current; (*b*) fluctuating direct current. 99
53. Positions of a coil rotating in a magnetic field. 99
54. Sine wave generated by the coil in Fig. 53. 100
55. The terms used to describe an alternating wave. 100
56. The current averaged over a half cycle of an alternating wave. 101
57. The principle of using vectors to find a resultant. 104
58. A rotating phasor tracing out a full sine waveform. 105
59. The summation of two sine waves point by point. 106
60. The phasor method of summing two sine waves. 106
61. An example of waveforms (*a*) in phase; and (*b*) out of phase by 90°. 108
62. Rectification showing: (*a*) original a.c. wave; and (*b*) rectified wave. 109
63. The simple principle of rectification. 110
64. A simple half-wave rectifier circuit and its operation over each half cycle. 111

65. (*a*) generated; and (*b*) load currents in a half-wave rectifier circuit. 112
66. (*a*) generated; and (*b*) load currents in a full-wave rectifier circuit. 113
67. A full-wave rectifier circuit. 114
68. A simple smoothing circuit. 114
69. Phasor representation of the two waveforms shown in Fig. 61. 115
70. A.c. supply connected to a pure resistance. 119
71. Phasor diagram showing voltage and current in phase in a purely resistive circuit. 119
72. A.c. supply connected to a pure inductance. 120
73. Phase relationship of current and voltage in a purely inductive circuit. 121
74. Phasor representation of the phase relationship between voltage and current in a purely inductive circuit. 121
75. Graphical representation of the phase relationship between voltage and current in a purely capacitive circuit. 122
76. Phasor representation of the phase relationship between voltage and current in a purely capacitive circuit. 122
77. Voltage and current waves in a capacitive reactive circuit. 123
78. Power wave for the capacitive reactive circuit shown in Fig. 77. 123
79. Graphical representation of the relationship between inductive reactance and frequency. 124
80. Graphical representation of the relationship between capacitive reactance and frequency. 125
81. Equivalent circuit for a coil having inductance and resistance. 127
82. Phasor diagram for the circuit in Fig. 81. 128
83. Phasor diagram for a series resistive-capacitive circuit. 128
84. Voltage triangle for Fig. 82. 129
85. Impedance triangle for a resistive-inductive circuit. 130
86. Phasor diagram for an LCR circuit. 132
87. Phasor diagram for a series resonant circuit. 134
88. Power triangle for an inductive circuit. 137
89. Voltmeter and ammeter connections showing ammeter error. 142
90. Voltmeter and ammeter connections showing voltmeter error. 143

91. Use of a mirror to obtain accurate meter readings. 144
92. Resistor used as a shunt to extend the range of an ammeter. 144
93. Resistor used as a multiplier to extend the range of a voltmeter. 145
94. Basic moving-coil galvanometer. 145
95. Basic principle of the repulsion type moving-iron meter. 146
96. Repulsion type moving-iron meter. 147
97. Typical scale for a moving-iron meter showing closeness of graduations for low readings. 148
98. Scale for a moving-iron with shaped iron rods. 149
99. The Wheatstone bridge. 151
100. Cells connected in series: (*a*) aiding; and (*b*) opposing. 153
101. Uniform wire *P* connected to a simple cell. 153
102. Simple method of null measurement using a potentiometer. 154
103. Resistance measurement by means of a potentiometer. 155
104. A commercial multimeter showing a typical arrangement of scales. 157
105. Basic atomic structures of: (*a*) hydrogen; (*b*) helium; and (*c*) sodium. 161
106. Simple covalent bonding. 166
107. Structure of a typical n-type semiconductor material. 169
108. Structure of a typical p-type semiconductor material. 170
109. The buildup of potential at a pn junction. 171
110. The forward biased pn junction. 172
111. The forward and reverse characteristics of semiconductor diodes. 173
112. The British Standard symbol for the semiconductor diode. 173
113. The make-up of the alloy type of transistor. 174
114. The British Standard symbols for: (*a*) npn; and (*b*) pnp transistors. 175
115. Electron and hole migration in an npn transistor. 176
116. Electron and hole migration in a pnp transisotr. 177
117. The three basic modes of connection for a transistor: (*a*) common base mode; (*b*) common emitter mode; and (*c*) common collector mode. 178
118. Voltage distribution for (*a*) silicon and (*b*) germanium; biasing. 179
119. Circuit suitable for obtaining the: (*a*) common base

characteristics; and (*b*) common emitter characteristics. 180

120. Input characteristics: (*a*) common base mode; (*b*) common emitter mode. 181

121. Output characteristics: (*a*) common base mode; (*b*) common emitter mode. 182

122. Transfer characteristics: (*a*) common base mode; (*b*) common emitter mode. 184

CHAPTER ONE

Units and Circuits

> CHAPTER OBJECTIVES
>
> After studying this chapter you should be able to:
> * understand the basic units of mechanical and electrical science, and appreciate the relationship between them;
> * manipulate figures using the recommended SI multiples and sub-multiples;
> * apply Ohm's law, the superposition theorem and Kirchhoff's laws to series and parallel circuits involving one or more d.c. sources.

INTRODUCTION

When any science is being studied, the units in which quantities are measured are the building bricks. There is no way in which you are going to build a successful house, unless you know the size and shape of your building bricks. A description of the units and how they fit into the system is, therefore, the very basis of study, and it is important that these units are clear in your mind before you attempt to progress further.

The first step is to decide on where to start, i.e. with those units on which others can be based. The whole of our system of units can be built up using the basic units of length, mass and time—metres, kilogrammes and seconds. The size of these units has first to be decided by international agreement, so that a man who has a metre of cloth in one country, knows it will be the same length as the man who has a metre of cloth in another country. When these basic units have been decided, all other units can be linked to them.

The unit of electrical current, the ampere, is also a basic unit. This helps us make the bridge between electrical and mechanical units, and is extremely useful to include in our basic package.

All other units which are derived from these fundamental starting units provide us with an absolute system. The units themselves are called absolute units.

BASIC ELECTRICAL UNITS

The electrical circuit is a particular means of transferring energy from one point to another. The energy is transmitted by the movement of electrically charged particles around physical conductors, through gases and certain types of liquid. When conductors are used they connect the source of the energy to the device which is to make use of it. This device is often called the *load*.

The charged particles, which are called electrons, although physically very small, still require a force to move them. This force is provided by the source and is called an electromotive force (e.m.f.). When the physical connections are in the form of metallic conductors, connected from point to point between all the individual parts, the complete system is called an *electrical circuit.*

The movement of charged particles in a circuit is known as an electric current (symbol I), and the unit of current is called the ampere (symbol A). The driving force (e.m.f.) produced by the source, is measured in volts (symbol V).

The quantity of electrical energy moved around the circuit will depend on the time the circuit is operating. It is known as the electric charge (symbol Q) and is measured in coulombs (symbol C). These are related by the expression $Q = I \times t$, (amperes $\times$ seconds). This is a very good example of the original principle. The coulomb is the product of two basic units.

The physical conductors and the load both present an opposition to the flow of current. Energy is used up in driving the electrons against this physical opposition. This means that the energy available for use at point 1, nearest to the source, is larger than that available at point 2 in Fig. 1. That is, the potential energy at point 1 is greater than that at point 2. A *potential*

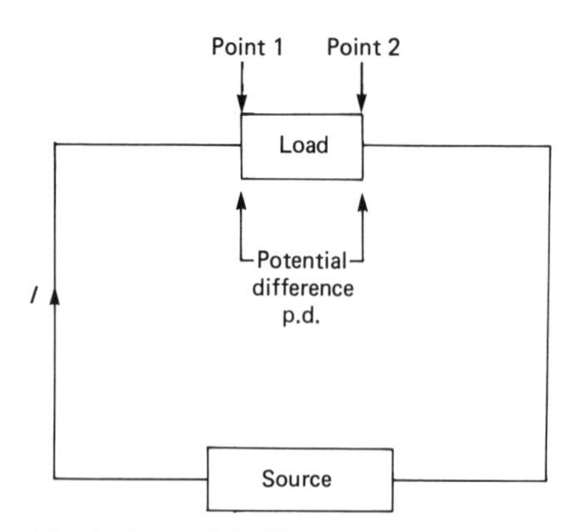

Fig. 1. *Potential difference across a load*

energy difference exists across the resistance. Since the original energy is measured in volts, this must also be measured in volts, but it is known as a potential difference or p.d. and not as e.m.f.

We can now see quite clearly that when all the voltages are measured around the full circuit, the original e.m.f. will be equal to the sum of all the potential differences. This is a very important principle of electric circuits, and applies throughout all branches of electrical engineering. It is the basis of Kirchhoff's second law which you will meet later.

The physical resistance (symbol R) is measured in ohms (symbol Ω).

An electrical engineer also needs to know how the electrical units which he uses regularly, link up with the mechanical units of force, work and energy. It is therefore useful to explain these at the same time.

MECHANICAL UNITS

The unit of mechanical force is the newton. When a mass sits on the earth, a force has to be applied to it in order to get it to move. A car will remain as a useless, stationary mass until a force (in this case produced by a petrol engine) is applied to it. When the force is applied, the car will accelerate, as long as the force is large enough to overcome the frictional resistance to movement. It is not too difficult to realise that the Formula One racing car has a bigger force applied to it than the little Mini, and hence it will accelerate better. An even bigger force will be applied to a fully armed tank, but its acceleration is relatively poor. There must be a link between force, mass and acceleration, and it can be defined using the other units. Newton's Second Law tells us that

$$\text{Force} = \text{mass} \times \text{acceleration}$$
$$(\text{newton}) = (\text{kgs}) \times (\text{metres/second}^2)$$

The newton is now quite easily defined. It is that force which will give a mass of 1 kg an acceleration of 1 metre/second^2.

The earth constantly exerts a force on all bodies near to it, tending to pull them towards its centre. The force pulling is the gravitational force and it is this which gives a mass its apparent weight. The unit in which it is measured is also the newton. However, the effect of the force diminishes as the distance from the earth increases. Thus an astronaut in space apparently becomes almost weightless, although his mass does not vary.

The Unit of Work

Let us go back to our car. The engine uses up energy in providing the force and eventually work is done in moving the physical mass of the car along the road. The amount of work done or energy expended, can conveniently be measured by accounting for the size of the force and the distance through which it acts. It is measured in newton-metres or joules.

$$\text{Work done} = \text{force} \times \text{distance}$$
$$\text{joules or newton-metres} = \text{newtons} \times \text{metres.}$$

The Unit of Power

Assume that you have to climb a long flight of stairs, say one hundred feet in height from top to bottom. You have to lift your body weight one hundred feet, and we can calculate the work done or energy expended in doing it. Now let us say that you do it twice, the first time in 45 seconds, but the next time it takes 90 seconds. The work you have done must be the same and some type of unit is required to demonstrate the difference in the obvious effort. The two climbs are at a different rate of expending energy or doing work and this is expressed as power (symbol P). Power is measured in watts (symbol W) and therefore must be defined as:

$$\text{watts} = \text{joules or newton-metres/second.}$$

The Unit of Torque

There are engineering situations where a great deal of force is applied and yet the body does not move in a particular direction. An example of this is in a well or pulley. Imagine you are winding up a bucket full of water in a simple well. The force is applied at the end of the handle at right angles to the radius of the circle traced out by the handle.

When a force is applied in such a way that it tends to produce rotation, it is called a torque. The size of the torque is the product of the force and the perpendicular distance from the point at which the force is applied to the point about which the body rotates. In the case of the well, this will be the length of the arm (*see* Fig. 2).

$$\begin{aligned}\text{Thus torque} &= \text{force} \times \text{radius of application}\\ &= \text{newtons} \times \text{metres}\\ &= \text{newton-metres (Nm)}\end{aligned}$$

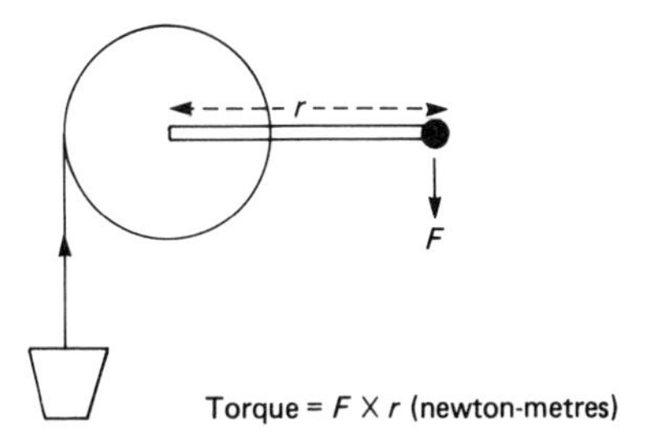

Fig. 2. *A simple illustration of torque*

Work done per revolution
The work done can easily be deduced by calculating the circumference of the circle traced out by the handle and applying the force × distance moved, expression.

MULTIPLES AND SUB-MULTIPLES OF UNITS

We have examined earlier what is meant by one metre of cloth, and we have an idea in our mind how far one metre stretches. A problem arises however if we want to describe a much larger distance, e.g. the length of a journey from Leeds to London. This is approximately 320,000 m, or alternatively, from London to Edinburgh, approximately 680,000 m. It obviously becomes inconvenient to measure and quote large distances in metres. The basic system is adapted to cope with the problem.

Multiples

It is obviously much easier to introduce another unit, which is simply a number of metres combined. It is convenient to decide that this second unit will be readily divisible for ease of calculations and in this case the well known kilometre meaning 1,000 m, is used. The two distances previously quoted are therefore much more sensibly expressed as 320 and 680 km respectively.

A second example is electrical resistances which are also of quite large values. 22,000 Ω is therefore expressed as 22 kΩ. Examining carefully these two examples, we can see that the basic unit still appears but in each case it is "prefixed" by the term "kilo". Indeed this is the method by which all multiples are described and below is shown the list of the preferred values.

Multiplication	Prefix
$\times$ 1,000 (10^{3})	kilo (k)
$\times$ 1,000,000 (10^{6})	Mega (M)
$\times$ 1,000,000,000,000 (10^{12})	Tera (T)

Sub-multiples

Similarly in some cases the basic unit is very large. A typical example of this is a capacitance which we shall meet later. The basic unit is the *farad*. This is far too large a unit for practical purposes. Much more common in engineering applications is the microfarad, which is the one millionth part of a farad. Even the picofarad which is the million-millionth part is a practical unit. In this case the opposite argument to that used for large distances applies and, expressed in farads, the value of 100 picofarads is 0.0000000001 F, which is very impractical. A range of sub-multiples is therefore employed and is shown in the table.

Multiplication	Prefix
$\times \frac{1}{1,000}$ (10^{-3})	milli (m)
$\times \frac{1}{1,000,000}$ (10^{-6})	micro (μ)
$\times \frac{1}{1,000,000,000,000}$ (10^{-12})	pico (p)

Conversion between Multiples and Sub-multiples

Although it is very convenient to express values in multiples and sub-multiples, it is sometimes necessary to convert from one form to the other. This often causes students considerable difficulty, but should not do so if a logical approach is applied.

When you are asked how many eggs you have in a bag, little difficulty is experienced if you have $2\frac{1}{2}$ dozen eggs. The unit of 1 dozen is known to contain 12 eggs and a simple *multiplication* is made, i.e. $12 \times 2\frac{1}{2}$ to find the total number of eggs. The dozen is the multiple unit. Hence to convert back to the basic unit, in this case 1 egg, multiplication must take place.

Exactly the opposite argument is true when it is required to express 500 milliamps in the basic unit of amperes. It is obvious that there is only a fraction of one whole ampere, or otherwise the sub-multiple would not be needed. The 500 must therefore be written as a fraction. A whole ampere contains 1,000 mA and the fraction will be:

$$500 \text{ mA} = 500/1{,}000 \text{ A}$$
$$= 5/10 \text{ or } 0.5 \text{ A}.$$

CIRCUITS

Many students have used Ohm's law, but if the truth were known it is probable that less than half have really understood it. The application of all rules must be logical, and the conditions must be correct for the rule to be applied in the first place. It is of little use applying a rule which says how quickly a room will warm up when a 2 kW electric fire is switched on if, at the same time, a door is left wide open and an icy draught is blowing through. The basic conditions must be correct first. Once these initial conditions have been correctly satisfied, the actual application is usually quite simple. Let us look at the factors which we must keep in our minds when the simple expression $I = E/R$ is being applied. Where E is the e.m.f.

The Four Important Conditions

Firstly, we must check the basic conditions which *always* hold good in all series and parallel circuits for d.c. application.

Series Circuits (*see* Fig. 3)

(*a*) In all sections of a series circuit the current must have a constant value.

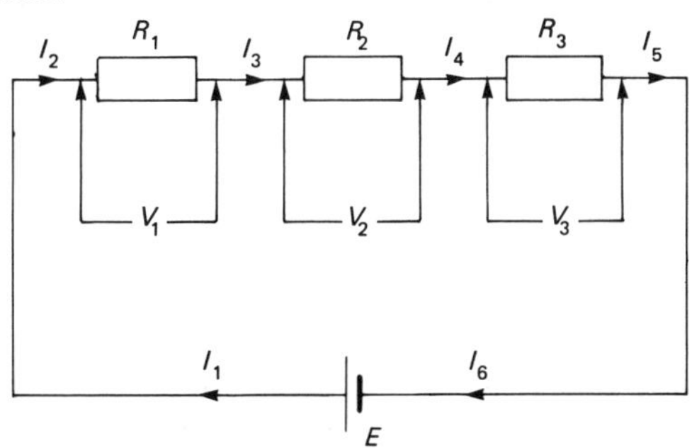

$I_1 = I_2 = I_3 = I_4 = I_5 = I_6 = I$
$E = V_1 + V_2 + V_3$

Fig. 3. *Voltage and current distribution in a series circuit*

(b) The potential difference across each component varies around the circuit, and is proportional to the d.c. resistance of each component.

Parallel Circuits (*see* Fig. 4)

(c) No matter how many components are connected in a parallel group, the potential difference across each of them is the same.

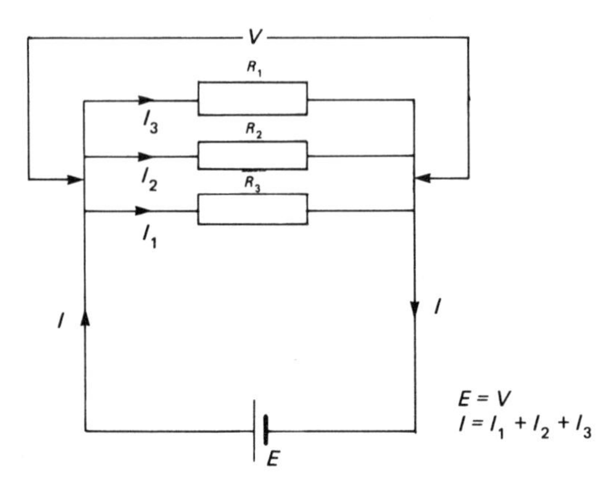

Fig. 4. *Voltage and current distribution in a parallel circuit*

(d) The current flowing through each of these parallel components will vary. The actual value of current will be inversely proportional to the value of its d.c. resistance.

These are the fundamental conditions and must always be applied when a circuit calculation has to be made. Figures 3 and 4 illustrate the four points listed.

Solving an Ohm's law problem is best achieved by following a set pattern, at least until a fuller understanding is gained. This is a logical approach and fewer mistakes are likely to be made.

Applications

Let us first consider the overall circuit and concentrate on finding the values of total resistance, total current and total voltage applied. Very often some of these values are given, particularly the applied voltage, but nevertheless concentrate on dealing with the whole circuit. The total resistance will often have to be deduced using the techniques which you have learned earlier. Quickly summarised these are:

in series circuits:

total equivalent resistance = Sum of all the individual resistance

$$R_T = R_1 + R_2 + R_3 \ldots, R_n;$$

in parallel circuits:

If the total equivalent resistance is given by R_T then

$$\frac{1}{R_T} = \frac{1}{R_1} + \frac{1}{R_2} + \frac{1}{R_3} + \ldots, \frac{1}{R_n}.$$

When these values have been obtained, concentration can be turned towards the individual components. Provided that the four basic conditions are remembered, the values of all currents and voltages around the circuit can be calculated using Ohm's law. The technique is best explained using an example.

Example
It is required to know all voltages and currents at every part of the circuit in Fig. 5.

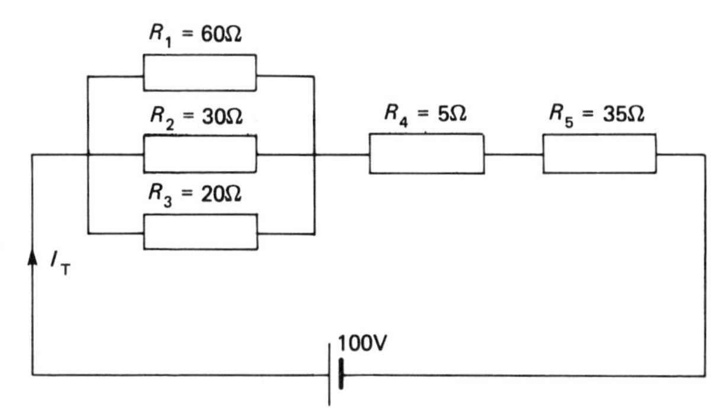

Fig. 5. *Circuit diagram for Ohm's law solution*

Always remember to set the problem out logically. This helps in two ways.

(*a*) It clarifies your own thinking.

(*b*) It helps an observer to follow your reasoning.

Remember the first thing to do is to treat the circuit as a whole. Total voltage must be the applied voltage = 100 V. Total resistance R_T – here we must apply the expressions quoted earlier to find the total equivalent resistance. We take the parallel group first. Let R_p be the equivalent parallel resistance.

$$\therefore \qquad \frac{1}{R_p} = \frac{1}{R_1} + \frac{1}{R_2} + \frac{1}{R_3}$$

$$\frac{1}{R_p} = \frac{1}{60} + \frac{1}{30} + \frac{1}{20}$$

$$\frac{1}{R_p} = \frac{1+2+3}{60}$$

$$\frac{1}{R_p} = \frac{6}{60}$$

$$\therefore \qquad R_p = \frac{60}{6} = 10\ \Omega$$

The parallel group has an equivalent resistance of 10 Ω. Now we can add this equivalent resistance as though it is in series with the other two resistors R_4 and R_5.

$$R_T = R_p + R_4 + R_5$$
$$= 10 + 5 + 35$$
$$R_T = 50\ \Omega$$

Next we have to find the total current flowing in the circuit, i.e. that taken from the d.c. supply. This must depend on the total applied e.m.f. and total resistance;

$$\text{and} \qquad I_T = \frac{E}{R_T}$$

$$= \frac{100}{50} = 2\ \text{A}.$$

We now know how the circuit over-all functions. Our attention can be turned to individual components in the circuit and we can determine the current flowing through each of them and hence the voltage drop, i.e. the potential difference across them. Remember the rules (*a*)–(*d*) which we listed earlier must now always be in the back of our minds.

Consider first, the resistors R_4 and R_5, since these are the least complicated. They are connected in series, and rule (*a*) says that the whole of the current supplied must flow through each series component. The answer can therefore be written down directly.

$$I_4 = I_5 = I_T = 2\ \text{A}.$$

We also know that whenever a current flows through a resistance a potential difference is produced across it. The size of this can be determined using Ohm's law, but the law must be applied to the individual resistance.

Hence
$$V_4 = I_4 \times R_4$$
$$V_4 = 2 \times 5$$
$$= 10 \text{ V}.$$

Similarly,
$$V_5 = I_5 \times R_5$$
$$= 2 \times 35$$
$$= 70 \text{ V}.$$

We now know all that is necessary about R_4 and R_5 and these can be left alone. Consideration must now be given to the parallel grouping. As far as the supply is concerned, it sees the equivalent resistance R_p as 10 Ω. The current divides through the parallel group, therefore the potential across the parallel group is given by

$$V_p = I_T \times R_p$$
$$= 2 \times 10$$
$$= 20 \text{ V}.$$

At this point we can check the validity of rule (*b*). This states that around the whole circuit the total e.m.f. will be equal to the sum of the potential differences across the components. Indeed we can see that this is true by using the following check.

$$E_T = V_p + V_4 + V_5$$
$$100 = 20 + 10 + 70$$
$$100 \text{ V} = 100 \text{ V}.$$

The next step must be to calculate how the current divides through the three resistance R_1, R_2 and R_3 of the parallel group. Here we have to keep rule (*c*) in mind. It says that all the resistors in a parallel group have the same potential across them. Ohm's law can be applied to each in turn to find the current.

$$V_p = V_1 = V_2 = V_3$$

$$\therefore \quad I_1 = \frac{V_1}{R_1} = \frac{20}{60} = \frac{1}{3} \text{ A},$$

$$I_2 = \frac{V_2}{R_2} = \frac{20}{30} = \frac{2}{3} \text{ A}$$

$$I_3 = \frac{V_3}{R_3} = \frac{20}{20} = 1 \text{ A}$$

Rule (*d*) says that for parallel groups the total current splits through the individual components. We can check our calculations are correct by applying this simple test.

$$I_T = I_1 + I_2 + I_3$$

$$2\ \text{A} = \frac{1}{3} + \frac{2}{3}\ \text{A} + 1\text{A}$$

$2\ \text{A} = 2\ \text{A}$, showing that our calculations are correct.

The details concerning all sections of the circuit are now resolved. On occasions it is necessary only to find the current or voltage associated with one component. This makes the solution shorter but the same basic principles always apply.

The only real method of becoming efficient at solving circuit problems is to practise. Other examples are given which you should attempt using the techniques shown.

THE SUPERPOSITION THEOREM

The circuits which have been considered so far have only had one source of e.m.f. in them. Provided the series of rules outlined in the previous paragraphs have been followed, the application of Ohm's law has been adequate to produce a solution. However, on many occasions in electrical and electronic engineering, more than one source of e.m.f. exists in a circuit. We must study the various techniques of extending Ohm's law to provide a solution in these cases.

One such technique is the superposition theorem. Although the name implies complication, its application is really common sense, provided a little care is taken.

Let us look at the circuit drawn in Fig. 6(*a*). Assume that a single cell is added at point A, having an e.m.f. of 2 V, and connected with the polarity shown in Fig. 6(*b*). An application of Ohm's law quickly produces the current shown of 0.2 A, which by convention, will flow in the direction of the arrow.

Now consider that the cell at A is removed and that a second cell is placed at B. In this case the battery has an e.m.f. of 5 V, and it is connected with the polarity shown in Fig. 6(*c*). This will produce a different current of 0.5 A in the opposite direction.

Let us now consider what the effect of adding both batteries at the same time will be. The circuit produced will be that in Fig. 6(*d*), and the total current will be the sum of the two individual currents taking into account their directions. The mathematician calls one direction positive and one negative. It is exactly the same electrically, and the total current will be $0.5 - 0.2 = 0.3$ A, in the direction shown in Fig. 6(*d*). When the mathematical signs are taken into account this is known as the *algebraic sum of the currents.*

We can now consider carefully what we have shown. What we have really demonstrated, is that when two sources of e.m.f. both attempt to influence the movement of electrons through the circuit, their combined effect is just the same as if we had considered each in turn, and then done the algebraic sum.

This is the basic principle of the superposition theorem. It says that if a circuit is complicated, because it has two or more sources

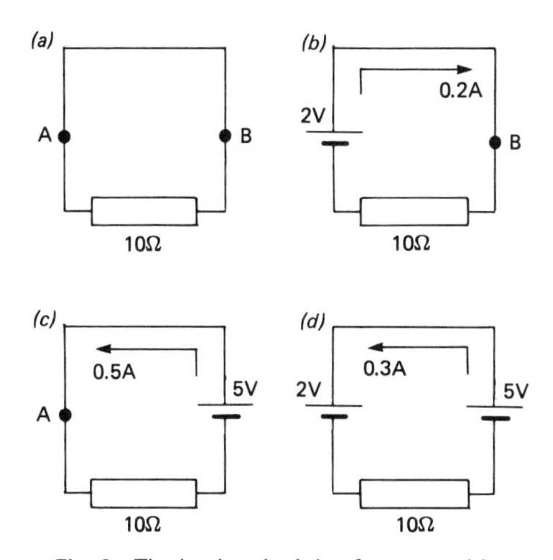

Fig. 6. *The basic principle of superposition*

of e.m.f., take each one in turn and calculate the currents which would flow due to that e.m.f. Then, after having calculated all the individual currents, the total current will be the algebraic sum of all the component currents.

One additional point must be remembered. The cell itself has a resistance, so that when the sources of e.m.f. are in turn neglected, they must be replaced in the circuit by their internal resistances or the behaviour of the circuit would change. In the early reasoning we ignored these internal resistances.

The theorem puts all this into words that sound rather complicated, but we will look at that later. Let us first do a simple example.

Example

Fig. 7(*a*) shows a circuit driven by two batteries of 10 V and 20 V respectively. The 2Ω and 4Ω resistors represent their internal resistances and are shown in series with the appropriate battery. We can apply the superposition theorem to determine the currents through the 10 Ω resistor.

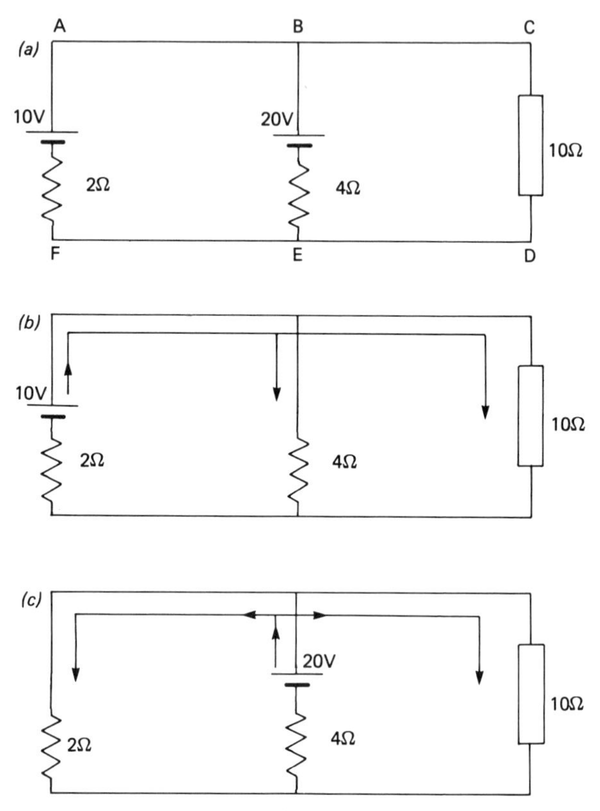

Fig. 7. *Circuit diagram for superposition solution*

We have to consider each battery in turn, replacing the other battery with its internal resistance before making the calculations.

Case 1

Replace the 20 V battery. Fig. 7(*b*) shows the modified circuit.

Applying Ohm's law consider the parallel group of 4 Ω and 10 Ω.

Let R_p be the equivalent parallel resistance

$$R_{p_1} = \left(\frac{1}{4} + \frac{1}{10}\right)$$

$$= \left(\frac{5+2}{20}\right)$$

$$\therefore \quad R_{p_1} = 20/7 = 2.86$$

$$\therefore \quad \text{total resistance } R_{T_1} = 2 + 2.86$$

$$= 4.86\ \Omega$$

$$\therefore \text{ total current} = \frac{E}{R_{T_1}} = \frac{10}{4.86} = 2.05\ \text{A}$$

Now apply Ohm's law to the individual resistor of 10 Ω. Not all of the current goes through the 10 Ω resistor and the ratio of split must be found. To find the current, we must first of all determine the voltage across the 4 Ω and 10 Ω parallel group. Their combined resistance, as we have already found, is 2.86 Ω.

Using Ohm's law:

$$V = I \times R_{p_1}$$

$$V = 2.05 \times 2.86$$

$$V = 5.88\ \text{V}.$$

Now in a parallel group all of the voltage appears across each component. Therefore the voltage across the 10 Ω resistor is 5.88 V. The current through it due to the 10 V battery must therefore be:

$$I_{10} = \frac{V_{10}}{R}$$

$$= \frac{5.88}{10} = 0.588\ \text{A}.$$

By convention the direction is as shown in the Fig. 7(*b*).

Case 2

This time replace the 10 V battery and replace it with its internal resistance. Fig. 7(*c*) shows the modified circuit. The total resistance offered to this battery is now different to that in the first case and:

$$\frac{1}{R_{P_2}} = \left(\frac{1}{2} + \frac{1}{10}\right)$$

$$= \left(\frac{6}{10}\right)$$

hence $$R_{P_2} = \frac{10}{6} = 1.67$$

and total $$R_{T_2} = 4 + 1.67$$

$$= 5.67\ \Omega.$$

The total current taken from the 20 V battery is therefore:

$$I_T = \frac{E}{R_{T_2}} = \frac{20}{5.67} = 3.53\ \text{A}.$$

Once again the total current does not all flow through the 10 Ω resistor. Part goes through the 2 Ω resistor, and part through the 10 Ω resistor. The ratio of the split must be calculated.

In this case it is the 2 Ω and 10 Ω resistors which are in parallel, and the voltage across this group must be calculated first. Their combined resistance has already been calculated as 1.67 Ω.

$$V = I \times R_{P_2}$$

$$= 3.53 \times 1.67$$

$$= 5.89\ \text{V}.$$

Now the current through the 10 Ω resistor can be calculated, because once again 5.89 V will appear across each component.

$$I_{10} = \frac{V}{R}$$

$$= \frac{5.89}{10} = 0.589\ \text{A}.$$

We have now found the current flowing in the 10 Ω resistor due to each battery in turn. The final answer is obtained by taking the algebraic sum.

Current due to Case 1 = 0.588 A.
Current due to Case 2 = 0.589 A.

Total current, since they are both in the same direction, is:

$$0.588 + 0.589 = 1.18\text{ A}.$$

This process is quite lengthy and many mistakes can be made in it. It is somewhat overcomplex for the simple circuit which we have used as an example. Other techniques such as the Kirchhoff method described later may be more suitable. Nevertheless, it is a very important electrical principle, and its application can very often be time-saving, particularly when several a.c. generators are the sources of e.m.f.

The Superposition Theorem

This can be stated simply as follows. When a circuit has more than one source of e.m.f., the current at any point in the circuit can be calculated by considering each source of e.m.f. separately, replacing all other sources by their internal resistances, then algebraically summing all the separate currents due to the different sources of e.m.f.

Students should note that for more advanced applications this simple definition will require some modification.

KIRCHHOFF'S LAWS

Really, we have already considered Kirchhoff's laws when applying Ohm's law to various types of circuits. However, it is the method of application to circuits which have more than one source of e.m.f. which must be outlined in detail if effective use is to be made of this important technique.

We cannot store electrons at points in connecting wires. Therefore, we know that the current at all points in a series circuit is the same. We also know that in a parallel circuit, the current splits at the junction. It is quite reasonable then to say, that if the electrons can not be stored up, however many billions of them come up to the junction, the same number must leave it. Kirchhoff's first law states just that.

Law 1

The total current reaching a junction must be equal to the current leaving it. This is sometimes expressed by saying that the algebraic sum of the currents is zero at the junction.

This can be shown in a simple example (*see* Fig. 8).

If the portion I_1 takes the upper branch, then the lower branch must have the original less I_1. This method of labelling the

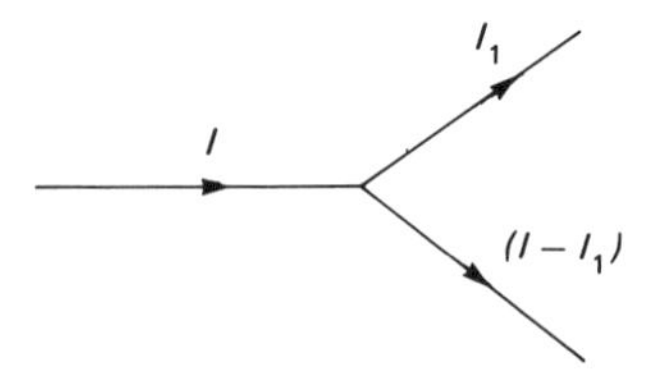

Fig. 8. *The principle of Kirchhoff's first law*

currents at a junction has the added advantage of using only two symbols. It will be seen later that this is important.

When we have been calculating series voltages around a circuit, we have learned that the sum of the potential differences across components will eventually add up to the total e.m.f. applied. This is Kirchhoff's second law.

Law 2

Around any closed circuit loop, the sum of the p.d.s is equal to the sum of the applied e.m.f.s.

Using these two laws, equations can be deduced for the circuit in such a way that the individual currents and voltages can be calculated. The most convenient explanation of this application is to use the same example previously solved using the superposition method.

The application of Kirchhoff's law should be taken a step at a time and completed to a set pattern, so that possible mistakes are avoided.

Step 1

Mathematical signs are very important in the application of Kirchhoff's laws. First choose a direction of current flow to be positive, mark it on the diagram, and keep to it. In this example we have decided that clockwise is positive (*see* Fig. 9).

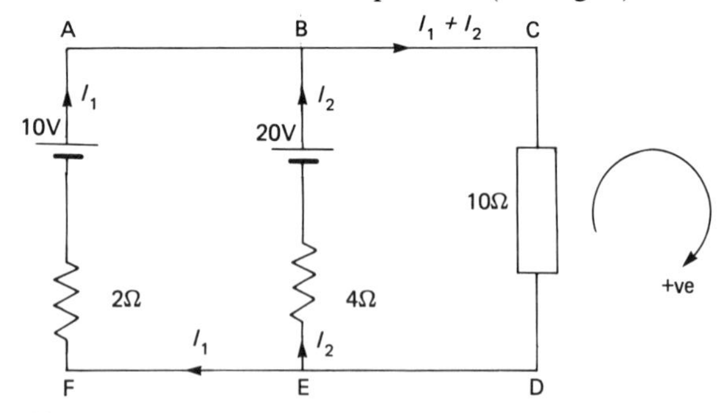

Fig. 9. *Circuit diagram for solution of problem involving Kirchhoff's law*

Step 2
Mark in all the currents flowing, remembering the first law and using as few symbols as possible.

Step 3
Decide on closed loops or as they are sometimes called "meshes". A closed loop is simply one on which you can put your pencil and trace all the way around back to the original point, e.g. ACDF or BCDE in our Fig. 9.

Step 4
Using only these closed loops and nothing else, we can apply the second law by equating the total e.m.f.s and the total p.d.s. Remember that a p.d. always exists across a component when a current flows through it.

In mesh ACDF the equation formed will be:

$$+ 10 = + 10\,(I_1 + I_2) + 2I_1. \tag{1}$$

Let us now look at this equation in detail and see how it has been formed. The left hand side of the equation is the sum of all the e.m.f.s around the ACDF loop. Trace it out and you will find that only one e.m.f. exists, i.e. the 10 V battery. Also, since by convention the current flows from the positive plate, the push of the e.m.f. is clockwise, and we have established that clockwise is positive. Therefore the left hand side of the equation is + 10 V.

Moving round the mesh, we next meet the 10 Ω resistor through which, by the application of the first law, $(I_1 + I_2)$ is the total current. Hence from $V = I \times R$ the p.d. will be:

$$\text{p.d.} = (I_1 + I_2) \times 10$$

This goes into the equation. Now since the direction of the current $(I_1 + I_2)$ is in the same direction as the one chosen to be positive, it is therefore $+\ 10(I_1 + I_2)$

Continuing through points D and F the next resistor is the 2 Ω. However, at this point the total current is only I_1. The I_2 portion has been diverted at the junction *E*. Hence from p.d. $= I \times R$:

$$\text{p.d.} = I_1 \times 2$$

This is the second term in the right hand side of the equation. No matter how complex the circuit is, provided this simple process is followed, the application of the Kirchhoff law is quite straightforward. Again the direction of I_1 will make it positive.

Remember, that provided the equations are formed correctly the problem is almost solved. It is very important to practise this technique of setting up the equations carefully. Now try to form the equation for the mesh BCDE for yourself by following the above procedure.

You should have produced the following result for mesh BCDE.

$$+20 = +10\,(I_1 + I_2) + 4I_2 \qquad (2)$$

So far so good, but all the parts of the expression have been positive so far. Let us now consider the mesh ABEF. There are two sources of e.m.f. in this case, of 10 V and 20 V. The left hand side of the equation will therefore read as follows:

$$+\,10 - 20\text{ V}.$$

The 10 V battery, by convention, will drive a current from the positive plate around the circuit in the direction which we have chosen to be positive, hence + 10 V. However, the 20 V battery will attempt to drive a current in the opposite direction to this and since it opposes our chosen positive direction, it must be labelled negative, hence − 20 V.

The right hand side will also have two components $2I_1$ and $4I_2$. Once again the current through the 2 Ω resistor flows in our positive direction and a current of $+\,2I_1$ will result. This is not so in the case of the 4 Ω resistor. Remember we are only considering the closed mesh ABEF, and within that mesh, $4I_2$ would be moving anticlockwise and therefore $4I_2$ will have a negative sign.

The whole equation is:

$$+\,10 - 20 = +\,2I_1 - 4I_2. \qquad (3)$$

There are only two unknown symbols marked in the circuits I_1 and I_2, therefore any two of these three equations, all of which are accurate for the circuit, are required to solve the problem.

Step 5

The fifth step is therefore to choose two and solve the pair of simultaneous equations to obtain the required values of I_1 and I_2.

We will now do this for the example given.

$$10 = 10(I_1 + I_2) + 2I_2 \qquad (1)$$

$$20 = 10(I_1 + I_2) + 4I_2 \qquad (2)$$

Hence

$$10 = 12I_1 + 10I_2 \qquad \times 10$$

$$20 = 10I_1 + 14I_2 \qquad \times 12$$

We solve this using the normal methods: by manipulating the equations to give one unknown which will have the same value. In this case we can multiply equation (1) by 10 and equation (2) by 12.

Thus
$$100 = 120I_1 + 100I_2 \quad (3)$$
$$240 = 120I_1 + 168I_2. \quad (4)$$

Subtracting equation (3) from equation (4)
$$140 = 68I_2$$
$$\therefore \quad \frac{140}{68} = I_2 = +2.05 \text{ A.}$$

The current I_1 can be calculated by substituting the value for I_2 back into an equation, in this case equation (1).
$$10 = 12I_1 + 10I_2 \quad (1)$$

Substitute
$$10 = 12I_1 + 20.5$$
$$10 - 20.5 = 12I_1$$
$$-10.5 = 12I_1$$
$$\frac{-10.5}{12} = I_1 = -0.875 \text{ A.}$$

Now the current which flows through the 10 Ω resistor is $I_1 + I_2$ and this is + 2.05 − 0.875.
$$\therefore \quad I_{10} = + 1.175 \text{ A.}$$

This is the same answer as obtained by superposition.

Step 6

This final step is very important. We made the original assumptions that the currents I_1 and I_2 flowed in a certain direction and called those directions positive. We must now make a final analysis and see whether or not our assumption was correct. The answers which we have obtained show I_2 positive but I_1 negative. Therefore, what is really happening in the circuit is that the 20 V battery is driving a current through the 10 Ω resistor *and* the 10 V battery. This final analysis is therefore essential. Fig. 10 shows the true conditions.

At first the procedure appears involved, but persevere, and quite quickly the ability to solve complicated problems will come to you. Further problems, which you should practice carefully, are given at the end of the section.

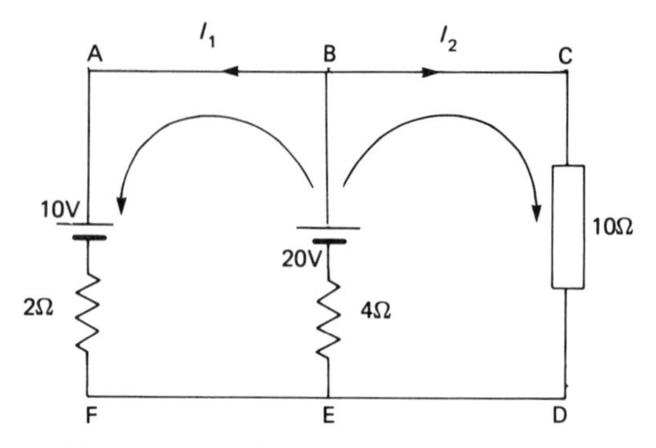

Fig. 10. *Final solution of problem involving Kirchhoff's law*

SELF-ASSESSMENT QUESTIONS

1. Convert 320 mA to amperes.
2. How many microfarads are there in 0.006 farads?
3. Express 4680 μC as coulombs.
4. The current through each component in a series circuit:
   (*a*) is always double the voltage applied;
   (*b*) has the same value;
   (*c*) splits proportionally to the value of the resistors;
   (*d*) splits inversely to the value of the resistors.
5. Kirchhoffs second law states that:
   (*a*) in any closed loop the sum of the current is zero;
   (*b*) the current through each component is always constant;
   (*c*) around any closed loop the sum of the e.m.f.s applied is equal to the p.d.s;
   (*d*) the algebraic sum of all the p.d.s at a point is zero;

6. Three coils have d.c. resistances of 8 Ω, 12 Ω, 24 Ω. They are connected (*a*) in series, then (*b*) in parallel. Calculate the voltage across each coil and the current through each coil for (*a*) and (*b*) if the applied voltage is 44 V.

7. Using Kirchhoff's laws calculate the current in the 12 Ω resistor.

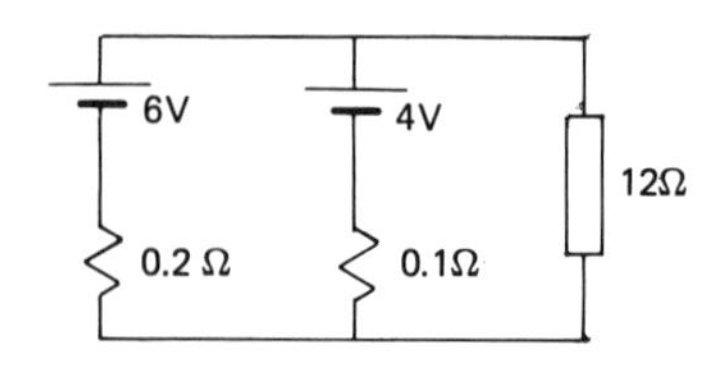

8. Using first Kirchhoff and then superposition find the current through the 8 Ω resistance in the following circuit.

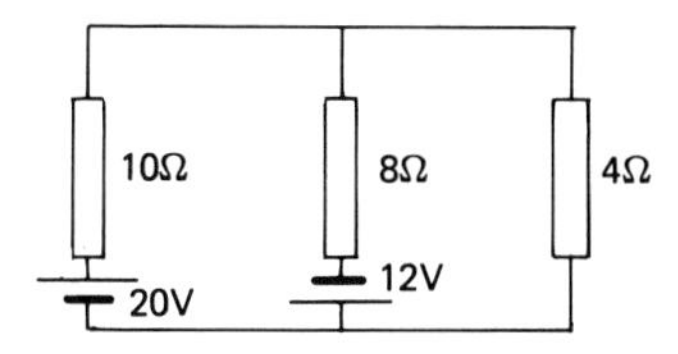

CHAPTER TWO

Magnetic Fields, Circuits and Effects

CHAPTER OBJECTIVES

After studying this chapter you should be able to:

* understand the basic units associated with magnetic circuits;
* appreciate the effect of ferromagnetic materials on the circuit performance;
* understand hysteresis, its effects and the terms associated with it;
* calculate the reluctance, total flux and flux density for a given magnetic circuit of known dimension and properties;
* appreciate the differences and similarities between electrical and magnetic circuits.

REVISION OF BASIC PRINCIPLES

It is important to look again at the elementary principles of the simple electric circuit before trying to study the magnetic circuit. Firstly, the circuit must be continuous, each element being connected to the next by physical conductors. Secondly, an electromotive force (e.m.f.) must be applied to the circuit in some form, a simple cell or battery. This can be considered as the prime mover in the system which, provided the circuit is continuous, will give rise to an electric current.

We have already learned that a current is the movement of charged particles around a continuous path. This current can be made to perform many useful functions in everyday life, such as heating and lighting. The current can be considered as the end product in the circuit. The size of the end product is important to the engineer and will be determined by the physical factors associated with the circuit, such as the size of the conductors, their material and their temperature. These physical properties oppose the movement of the charged particles around the circuit and are said to present resistance to the current flow. Naturally, the three basic quantities have all to be measured, and units have been allocated to them.

The end product or current is measured in *amperes*.
The prime mover or e.m.f. is measured in *volts*.
The opposition or resistance is measured in *ohms*.

The units should be familiar to you by now, but it is very important to fully understand these basic ideas about the electrical circuit before attempting to study the new idea of a magnetic circuit. The flow of an electric current can give rise to three major effects.

(*a*) *Heating*. This effect is widely used in domestic water heater tanks, electric fires, etc.

(*b*) *Chemical*. The passage of a current can give rise to chemical changes, which may have a useful application, for example electroplating for chrome or silver. Conversely, a chemical change can provide a source of e.m.f. such as a torch or car battery.

(*c*) *A magnetic effect*. This is the effect which, in this chapter, we particularly want to consider.

When a current flows along a conductor a magnetic field is established around it. Properly applied, this simple effect can be used to give many advantages in our everyday lives. The electric motor for public transport in vehicles or lifts, relays which control circuits for telephones or automatic operations and the simple door bell or buzzer all rely on this fundamental magnetic effect. The principles behind these applications will be dealt with in more detail later but first we will consider the basic magnetic field.

SIMPLE ELECTROMAGNETIC FIELDS

The shape of the field set up around a single conductor is shown in Fig. 11. It has direction, which depends upon the direction of current flowing in the conductor. A simple rule helps us determine this direction. Imagine a screw being screwed along

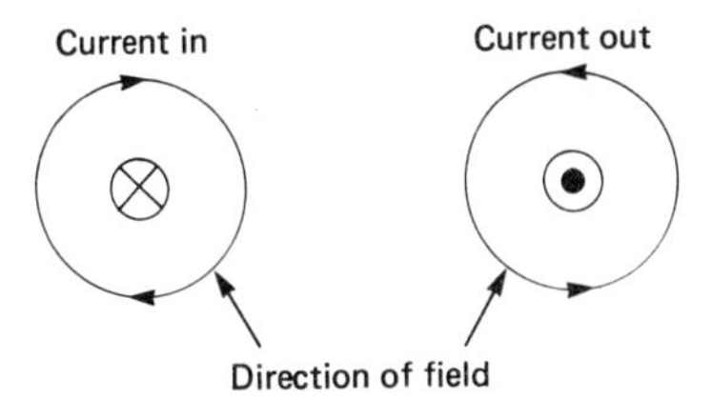

Fig. 11. *Magnetic field around a single conductor*

the wire in the direction of the current. The direction of rotation of the screwing gives the direction of the field.

The cross and dot shown are a simple way of indicating which direction the current is flowing when a cross section of the conductor is considered. Imagine an arrow along the centre of the conductor, pointing in the direction of the current. When the current is moving away from you, you would see the flight, and a cross is shown. When coming towards you, you would see the point of the arrow and hence a dot is shown.

The field around a single conductor is not in itself very useful to us, because it is likely to be weak and is only effective near to the wire. However, when the single conductor is formed into a coil, often called a solenoid, the many turns of wire set up a

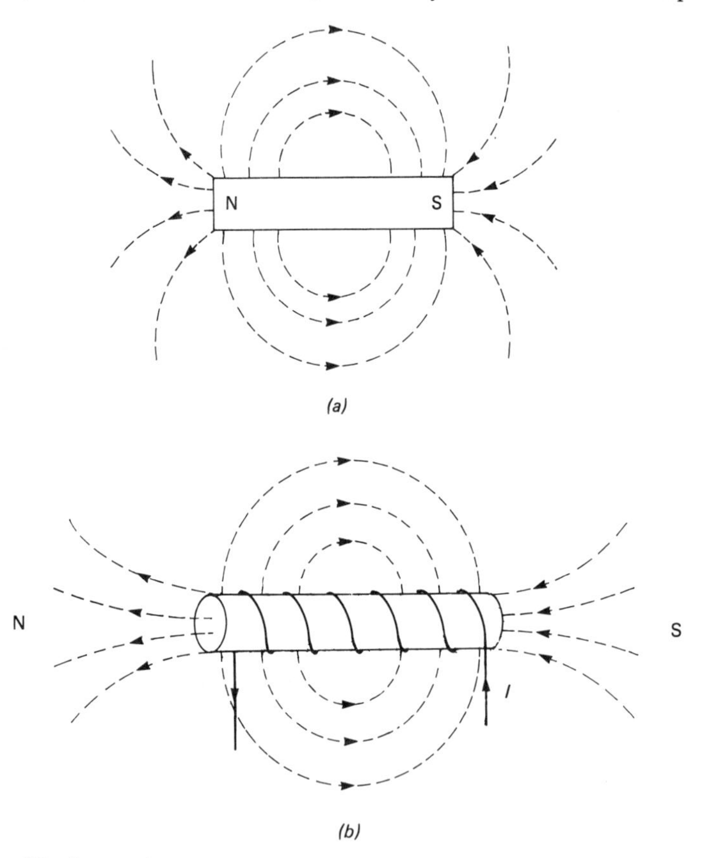

Fig. 12. *Comparison of the Magnetic fields around* (a) *a bar magnet; and* (b) *a solenoid*

much stronger field and also polarise the coil. You will remember from earlier studies, that the simple permanent bar magnet has a north and a south pole. The solenoid behaves in a similar manner when a current is passed through it. A field pattern very similar to the bar magnet is set up and poles appear at the ends of the coil, which have the same properties as those of the bar magnet. A comparison is shown in Fig. 12.

The field pattern for the individual conductors combine and produce an overall pattern which can be seen is almost identical to that of the bar magnet. The resultant pattern is very much more useful than that of the single conductor. It is a simple matter to wind a coil of copper wire, pass a current through it and produce a magnet which displays all the properties of the permanent magnet. This magnet depends upon the current flowing through it and hence is called an *electromagnet.*

We now have to examine the properties of such a magnet in more detail. This effect seems to suggest that as the current through the coil changes, the properties of the magnet change. Since a single turn of wire has only a weak magnetic effect and many turns produce a useful magnet, the number of turns of wire must have a direct influence on the properties of the electromagnet. Using a simple coil we can now examine how these factors vary in detail.

This magnetic circuit has similarities to the electric circuit but it also has some different properties. These properties are expressed by new words, which you have probably not met before, and it is important to get these clear in the mind in order to understand fully the working of the simple magnetic circuit.

THE RELATIONSHIP BETWEEN SIMPLE MAGNETIC UNITS

Magneto-motive Force

There has to be a prime mover in any system to make things happen. In the magnetic circuit it is known as the "*magneto-motive force*". This is usually abbreviated to m.m.f. We have already considered two factors which may well increase the effect of the prime mover in the circuit i.e. the current and the number of turns. This is exactly right and the magneto-motive force simply reflects these two properties. Hence:

magneto-motive force (m.m.f.) = current × number of turns.

We know that the basic unit of current is the ampere and the number of turns is just a pure number. A logical unit in which to measure m.m.f., is the *ampere-turn*. However, the SI unit for m.m.f. is the ampere but this may cause confusion at this stage and it is recommended that the unit ampere-turn be used. Thus:

m.m.f. (ampere-turns) = current (amperes) × number of turns (turns).

Magnetising Force (*H*)

Consider this very simple question. When an athlete runs around a track he performs an amount of work, i.e. he expends energy. How much energy will he expend? This is impossible to answer without knowing the length of the track. Obviously if the track is 100 m long he will use less energy than if the track is 1 km long.

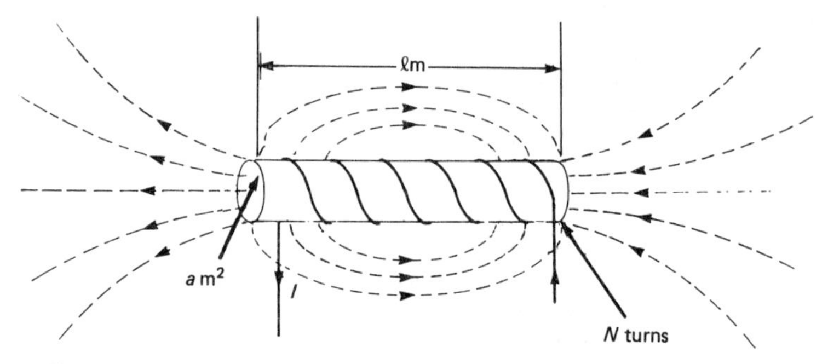

Fig. 13. *A simple coil electromagnet, showing its important dimensions*

There is a similar problem with the magneto-motive force. There may well be 10,000 turns and a large current of, say 2 or 3 A, but if the prime mover is applied over a length 10 m, it is likely that it will have a very different effect than when it is applied over, say, a centimetre. A second term has, therefore, been included in the "magnetic language" designed to take account of this very fact. This term is called the *magnetising force* and compares the magneto-motive force with the length over which it is applied. Looking at Fig. 13:

$$\text{Magnetising force } (\boldsymbol{H}) = \frac{\text{magneto-motive force}}{\text{length of coil}} = \frac{\text{ampere-turns}}{\text{metres}}$$

$$\boldsymbol{H} = \frac{I \times N}{l} \text{ At/m}$$

The unit ampere-turns/metre has to be learned, but it is common sense if the argument is followed through carefully. Note that the symbol for magnetising force is ***H***.

Flux and Flux Density

The lines of force shown around a single conductor in Fig. 11 or around a coil in Fig. 12 represent the magnetic flux. In the case of the magnetic circuit, this is the end product, this is what we are trying to achieve. In the electric circuit the e.m.f. is the energy source which causes a current to flow around a circuit. In comparison, in a magnetic circuit the m.m.f. is the energy source which establishes the magnetic flux. It is this magnetic flux, and particularly the strength of it, which we can use for many purposes in everyday life.

The total flux can be measured and is expressed in Webers (Wb). The Greek letter Φ is used as the symbol for the total flux. However, the same sort of problem exists as did with magnetomotive force. The flux is set up in three dimensions and hence saying that a circuit has a total flux of x Wb does not really indicate how strong a magnetic field we have managed to produce.

Ask another simple question. If there are 150 people in a room, how crowded are they? Again, this is impossible to answer without knowing the area of the room. Once this is known the concentration of bodies can be deduced as in units of people per unit area.

Similarly, a much more useful measure of the effectiveness of the magnetic flux is the amount of flux per unit area. This tells how much flux is concentrated in a given area and indicates the overall strength of the magnetic field. It is called the *flux density*. In this case the symbol $\boldsymbol{B}$ is used.

Referring back again to Fig. 13, we can see the relationship between the two quantities, flux Φ and flux density $\boldsymbol{B}$. When the flux is Φ Wb and the cross section area of the coil through which all the flux passes is a m^2, the flux density is defined as:

$$\text{flux density } (\boldsymbol{B}) = \frac{\text{total flux } \Phi \text{ (Wb)}}{\text{cross sectional area } a \text{ (m}^2\text{)}} .$$

$$\text{Effective unit} = \text{Wb/m}^2.$$

The actual unit used for flux density is the tesla.

Permeability μ

We would expect that for a current I, a coil would produce a field of a particular flux density. This will be established whenever the current is switched on. The relationship between the flux density $\boldsymbol{B}$, and the magnetising force $\boldsymbol{H}$ will remain a constant, provided other factors do not change. The ratio $\boldsymbol{B}/\boldsymbol{H}$ really shows how successful you have been in setting up the flux, compared to how hard you tried, i.e. how big a magnetising force was applied. The ratio is given the name *permeability*. It shows how magnetisable is the medium through which the flux is passing and, as we shall later see, is a very good indicator as to how good a magnetic material we are using. The symbol used is another greek letter, μ, and the unit is the henry/metre.

Permeability of free space μ_0

Let us assume that in our circuit the coil is placed in a vacuum. A vacuum is not a good magnetic medium and if $\boldsymbol{B}$ and $\boldsymbol{H}$ are measured, and the ratio $\boldsymbol{B}/\boldsymbol{H}$ calculated, the resultant value of μ is $0.4\,\pi \times 10^{-6}\,\boldsymbol{H}$/m. This is an awkward figure to deal with, but its use is unavoidable. However, as we shall see later, it does provide a very useful standard against which other materials can be compared. The symbol for the permeability is therefore modified a little to indicate this standard and is called μ_0 represents the permeability of free space or of a vacuum. This indicates that the values of $\boldsymbol{B}$ and $\boldsymbol{H}$ have been compared when the coil was set in a vacuum.

Relative Permeability μ_r

Let us now consider what happens when we have various materials in the centre of the coil, e.g. wood, aluminium, hard steel, air and a soft iron. The coil, with a known number of turns, can be set up with each type of core in turn, and an identical current I passed around it. We can calculate the magnetising force $\boldsymbol{H}$, and this will be a constant independent of the material in the centre of the coil. However, if we measure the flux in each case, we obtain some interesting results. The wood, aluminium and air give almost identical results to that obtained for the vacuum. Indeed, the difference is so small that, for all practical purposes, it can be assumed to be the same. The results obtained by using the other two materials are very different. The flux Φ and hence the flux density $\boldsymbol{B}$, is found to have a much greater value. This suggests an important principle, that the choice of material for the core of the magnetic circuit can influence the efficiency of the magnetic circuit.

Now, we already know that since permeability is the ratio $\boldsymbol{B}/\boldsymbol{H}$, and in our experiment $\boldsymbol{H}$ has the same value each time, when the value of $\boldsymbol{B}$ increases, so will the value of μ. The permeability can thus be a direct indication of how efficient a particular material will be and is used by the engineer for just this purpose.

One snag still exists. Nature does not often provide easy sums and the figures can be quite awkward, as we have already noted (e.g. $\mu_0 = 0.4\,\pi \times 10^{-6}\,\boldsymbol{H}/\mathrm{m}$). We can thus use samples of every type of material as cores for our coil and calculate the value of $\boldsymbol{B}/\boldsymbol{H}$ for every one. We shall finish up with a large number of clumsy figures. A rather more simple approach can be adopted with a little thought. Let us accept the figure for the vacuum and, as we have said, for air. What is now important is how much more efficient does putting a soft iron core into the coil make our circuit, and derive an easily usable comparative figure. In the case of iron it may be, say, 500 times more efficient, a hard steel 1,000 times, and so on. Therefore, instead of having a large number of awkward figures, a more common-sense approach is to have a list of comparison values and use these as multiplying factors. To avoid

ELECTROMAGNETIC		ELECTRICAL	
TERM	SYMBOL	TERM	SYMBOL
Magnetic flux	Φ	Current	I
Magnetic flux density	B	Current density	σ
Magnetising force	H	Surface charge density	J
Magnetomotive force	m.m.f.	Electromotive force	e.m.f.
Permeability of free space $\mu_o = 0.4\pi \times 10^{-6}$ H/m	μ_o		
Relative permeability	μ_r		
Absolute permeability $\mu = \mu_o\mu_r = \frac{\text{Magnetic flux density}}{\text{Magnetising force}}$ i.e. $\mu_o\mu_r = \mu = \frac{B}{H}$	μ	Resistivity	ρ
Reluctance $\left\{\frac{l}{\mu a}\right\}$	S	Resistance $\left\{\rho\frac{l}{a}\right\}$	R
Core possessing reluctance to establishment of flux; m.m.f.; Flux		resistance; Current; e.m.f.	
m.m.f. = Flux × Reluctance		e.m.f.= Current × Resistance	

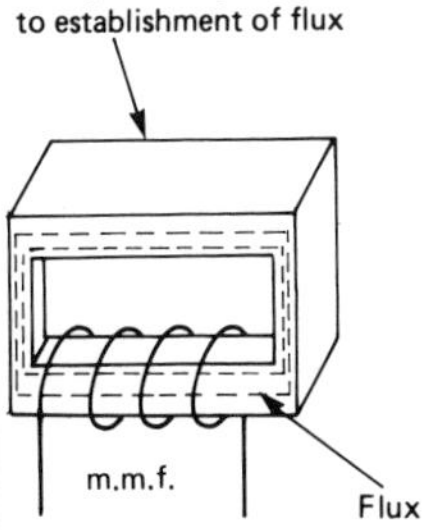

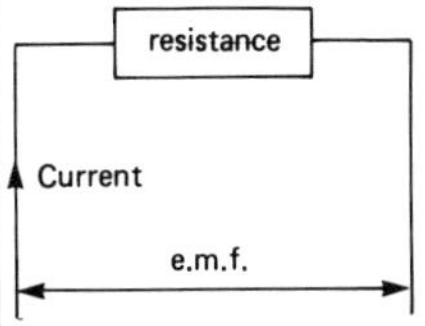

Fig. 14. *Comparison of electromagnetic and electrical terms*

confusion we have to distinguish it from the basic term "permeability". A logical choice is to call it *relative permeability*, since it indicates how much more efficient the particular material is, relative to a vacuum. The symbol has a suffix "r" and is shown μ_r. The total μ, called the absolute permeability of a material, is then a simple expression containing the two other terms.

Absolute permeability = permeability of free space × relative permeability

$$\mu = \mu_0 \times \mu_r$$

When experiments have been carried out and a full list compiled, a very important fact emerges. All the non-magnetic materials, including metals such as aluminium, brass etc., do not perform any better than air and hence all have the relative permeability value $\mu_r = 1$. This is very important and should be learned. A full comparison of electrical and magnetic terms is given in Fig. 14.

MAGNETISING CURVES FOR TYPICAL FERROMAGNETIC MATERIALS

In the experiment performed above we maintained the original set of conditions, i.e. that the current and number of turns remained constant. The values of μ were therefore constant for a given material. However, this is not always the case. When we gradually increase the value of I and hence the magnetising force $\boldsymbol{H}$ from zero, we find that for a time the flux density $\boldsymbol{B}$ is proportional to the increase in $\boldsymbol{H}$, i.e. μ is a constant. Eventually the increase in $\boldsymbol{B}$ starts to tail off and finally a further increase in $\boldsymbol{H}$ does not result in any further increase in $\boldsymbol{B}$. This is best seen in a simple graph relating $\boldsymbol{B}$ to $\boldsymbol{H}$ (Fig. 15).

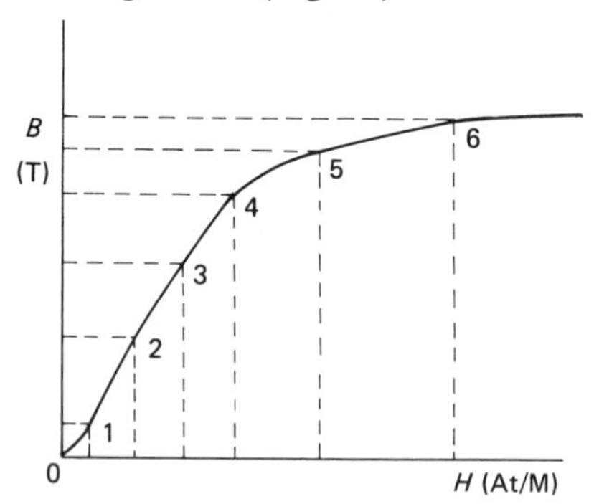

Fig. 15. *Changes in flux density* B *Resulting from changes in magnetising force* H

At the higher values of ***H*** the material is said to have reached saturation. It will be realised that the value of μ will therefore vary depending upon which part of the curve is being used, and the design engineer has to be aware of this when he designs his circuit. We can actually plot a graph showing how this value μ varies with ***H*** and since, in most practical cases, the number of turns on the coil remains constant it is really demonstrating how μ varies with the changing value of circuit current, which is what the design engineer really wants to know.

Referring back to Fig. 15, consider the six points marked. As we move from 1 to 2 the value of ***H*** has increased and the amount of the increase can be read from the ***H*** scale. The value of ***B*** will also have increased by a certain value and this can be read from the ***B*** scale. It is then a simple matter to calculate μ. Using the same procedure the value of μ can be calculated between points 3 and 4, and 5 and 6. A calculation shows that the values between 1 and 2 and 3 and 4 are the same, but between 5 and 6, since the increase in ***B*** decreases, so will the value of μ. Thus it can be deduced that μ is not a constant, but that it varies depending upon the value of the magnetising force ***H***.

Let us, for a moment, look at the principles associated with this varying value of μ. Firstly, since μ_0 is a constant, what we are really plotting are the values of μ_r against ***H***. The theory of the magnetisation process is very complex and we can only examine the general aspects involved. Nevertheless, we can get a picture of what is happening if we regard the magnetic material as being made up of a great number of very small individual magnets. In the basic material these are aligned in a completely random manner. Within the material, groups of these very small magnets tend to cluster together and have their magnetic axes all in the same direction. These groups are called domains. However, the overall effect of all the domains is that the resultant magnetic field is negligible. When a magnetising force is applied, those domains with axes very close to the direction of the field align themselves easily, and an initial magnetisation is quickly established. The value of μ thus rapidly increases during this period. The value of μ reaches its maximum at this point and it maintains this value over the portion of the ***B/H*** curve which is roughly linear. As ***H*** increases, more domains align and the increase in ***B*** remains proportional to the increase in ***H***. Eventually, when a large proportion of the domains become aligned, the rate of alignment decreases and the value of μ falls quite rapidly, in spite of a continued increase in ***H***. As saturation occurs, little further increase in ***B*** results and μ falls rapidly. The general shape of the μ***H*** curve is shown in Fig. 16.

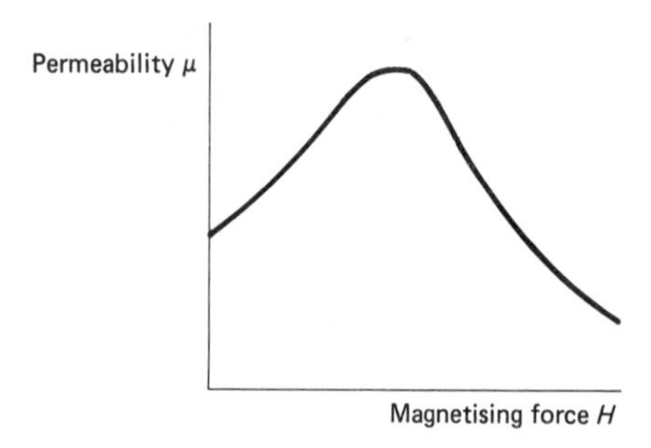

Fig. 16. *Typical curve showing variation in permeability with increasing* H

A typical range of values of some common magnetic materials is shown in the accompanying table in Fig. 17.

Material	Typical value of μ_r (max)
Cast steel	1,000
Cast iron	220
Wrought iron	2,700
Stalloy	6,500

Fig. 17. *Typical maximum μ_r values for common magnetic materials*

The Reluctance of a Magnetic Circuit (S)

We have already considered the effect of using a ferromagnetic core in the coil, compared to that of air or wood. It will be remembered that this considerably increased the efficiency of the electromagnet. The resultant flux increased far more rapidly for a given magnetising force, than was the case using an air core. Let us now look at it the other way around. Logically it would seem that the air or wooden core presents a greater opposition to the setting up of the magnetic flux, than does the ferromagnetic material. Here we have a third similarity with the electrical circuit. We have already compared the prime movers and the end products, now we must consider what effectively opposes the production of the end product. In the electrical circuit the opposition is called the resistance. In the magnetic circuit it is termed the reluctance. The symbol used is S. Also, just as the resistance is dependent upon the material and the dimensions of the conductor, so does the reluctance of a magnetic circuit depend upon the material and the dimensions. Let us briefly revise the expression for resistance.

$$R = \frac{\varrho l}{a} \Omega$$

where ϱ is the resistivity (Ωm), and depends upon the kind of material used;
l is the length (m) of the magnetic material;
a is the cross-sectional area (m^2) of the conductor.

A similar expression governs the reluctance of the magnetic circuit.

$$S = \frac{l}{\mu \, . \, a} \text{ (ampere-turns/weber)}$$

The dimensions play an important part and the value of S is relative to the material. Hence, the reluctance S can be calculated when these dimensions are known. The familiar expression for Ohm's law relates all these quantities in the electrical circuit, i.e.

$$\text{the end product} = \frac{\text{prime mover}}{\text{opposition}},$$

which is

$$\text{current } (I) = \frac{\text{e.m.f.}(E)}{\text{resistance}(R)}$$

It is not unreasonable to expect a similar relationship to occur in the magnetic circuit, and

$$\text{the flux } (\Phi) = \frac{\text{magneto-motive force (mmf)}}{\text{reluctance } (S)};$$

which means that the reluctance can also be expressed:

$$S = \frac{\text{mmf}}{\Phi}$$

and is logically given the unit ampere-turns/weber.

Let us apply this simple expression in a practical case.

Example
Calculate the value of the current flowing in a coil of 200 turns, if the cross-sectional area of the coil is 20 cm^2; the coil length is 50 cm; the relative permeability of the coil core is 1,500; and the flux produced is 0.001 Wb.

Finding the reluctance of the magnetic circuit;

$$S = \frac{l}{a\mu_0\mu_r}.$$

Using the correct units of length and area;

$$S = \frac{50 \times 10^{-2}}{0.4\,\pi \times 10^{-7} \times 1{,}500 \times 20 \times 10^{-4}}$$

$$\therefore \quad S = \frac{50 \times 10^{9}}{0.4\pi \times 1{,}500 \times 20}$$

$$= 1.326 \times 10^{6}\,\text{At/Wb.}$$

As $$S = \frac{\text{mmf}}{\Phi}$$

then $$\text{mmf} = S \times \Phi.$$

Hence, $$\text{mmf} = 1.326 \times 10^{6} \times 10^{-3}$$

$$= 1.326 \times 10^{3}\,\text{At.}$$

But $$\text{mmf} = I \times N.$$

Hence, $$I = \frac{\text{mmf}}{N}$$

$$= \frac{1.326 \times 10^{3}}{200} = 6.63\text{A.}$$

However, in a great many practical applications the magnetic flux has to be established across an air gap to be useful. Since we have already discovered that the air gap possesses considerably different magnetic properties than does a ferromagnetic material, this has a direct influence on the value of the current needed. The basic expression for reluctance remains the same, but of course in this case $\mu_r = 1$. The value of the reluctance for the air gap has to be separately calculated, and added to the value for the magnetic material. The flux has to permeate all sections of the complete magnetic circuit. To calculate this total opposition is straightforward, since it behaves identically to a simple electrical circuit. You will recall that the total resistance of an electrical circuit is calculated as follows.

$$R_T = R_1 + R_2 + R_3 + \ldots \text{etc.}$$

Similarly,

Total reluctance = the sum of the reluctance for the individual parts of the circuit, provided the flux is the same in all parts of the circuit.

$$S_T = S_1 + S_2 + S_3 + \ldots \text{etc.}$$

The circuit is called a simple series magnetic circuit and the similarity to the series electrical circuit is obvious.

Magnetic Screening

We have firmly established the fact that when a current flows in a conductor or coil a magnetic field is set up. The effect of this magnetic field is felt in the area surrounding the conductor, and depending upon its strength may well influence circuits a considerable distance away. This may give rise to several unwanted effects, some of which will be studied in the next chapter.

We have also found that magnetic materials are much more permeable than air, and wherever possible the flux will tend to concentrate in these materials rather than any non-magnetic alternatives. This gives us a very useful means of protection from unwanted effects of magnetic fields, if we wish to use it. Provided some highly permeable material surrounds the area which is to be protected, it can be virtually made free of field. The area is shielded from the effects and hence the term "magnetic shielding" or "screening" is used. Highly permeable materials have been developed specifically for this purpose, one such example being known as mu-metal. The principle is illustrated in Fig. 18.

Typical engineering examples of screening are in the protection of the magnetic recording head in tape recorders and the high-frequency transformers used in radio and television amplifiers.

Hysteresis

We have studied carefully in an earlier section how the magnetic flux density changes in a given specimen of magnetic material as the magnetising force changes. Eventually, the flux density reaches a maximum which we have termed saturation. There is no point in further increasing the current since no further increase in flux will be achieved.

Let us now consider the effect of steadily reducing the magnetising force. Logic would suggest that the flux density will diminish to the same values as were in evidence on the buildup of the field. However, this is not so, as can be seen in Fig. 19. The dashed line X shows the initial buildup of flux from the completely unmagnetised state. When the magnetising current is reduced from the point of saturation the curve Q is followed. Some of the magnetic domains are re-orientated and no longer contribute to the net magnetic polarity. As the value of $+\boldsymbol{H}$ is reduced still further, more and more domains disorientate and

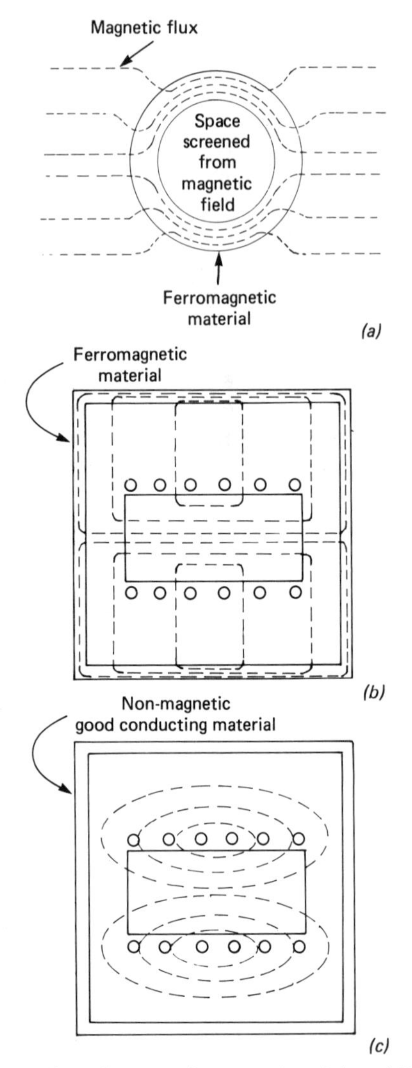

Fig. 18. *Three examples of magnetic screening:* (a) *and* (b) *low frequency currents;* (c) *high frequency currents*

the flux density is further reduced. However, when the value of ***H*** is zero, i.e. the current has ceased to flow, a very interesting effect is noticed. The flux has not been reduced to zero and the coil still behaves as an electromagnet which is weaker but still has the same polarity. This residual flux density is called the *remanent flux density X*. In this case, as saturation has been reached, the value of the remanent flux density is known as *remanance value for the material*. The amount of remanent flux density which remains depends on the type of material used in the core of the coil. Its value is shown at point R.

The terminology + ***H*** is shown on the graph simply to denote a magnetising force set up by a current in one particular direction. The term − ***H*** denotes a reversal of the current and the effect of doing this is what we must consider next.

The retained magnetism has a particular polarity, and a reversal of the current will attempt to produce an electromagnet

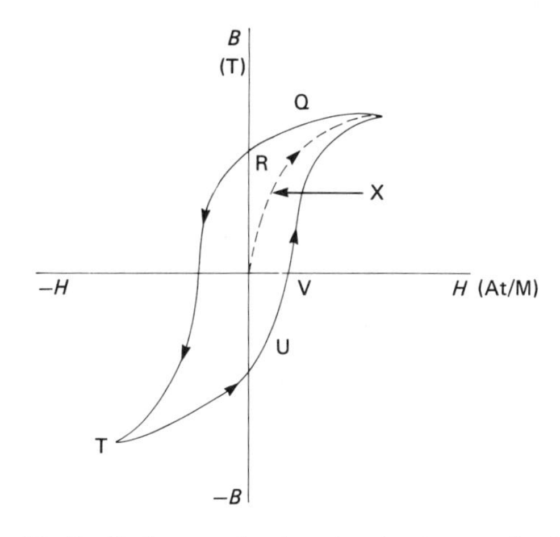

Fig. 19. *Cyclical magnetisation showing hysteresis effect*

of the opposite polarity. This means that the domains, which are still aligned in the original direction and which are causing the retained magnetism, will be subject to a force which causes them, once again, to take up random positions within the material. There will eventually be no overall magnetic effect. The amount of "negative" magnetising force which will reduce the flux density from the value at R to zero, is called the coercive force. In the case (after saturation has occurred) this is known as the *coercivity* of the material. As ***H*** continues to be increased in this reverse direction, the flux density will start to increase and will eventually reach saturation, producing the reverse polarity. Reduction of − ***H*** will produce a curve, through T and U, similar to that produced when reducing + ***H***. When the magnetising force has again been reduced to zero it can be increased in the positive sense, eventually bringing the flux density back to zero at the point V. Further increase in $+ \boldsymbol{H}$ will eventually result in saturation, thus completing the loop. Successively changing the value of $+ \boldsymbol{H}$ and $- \boldsymbol{H}$ will reproduce this loop repeatedly. Throughout, it will be observed that the flux density, once established, effectively resists the changes which are being made to it, and this gives rise to losses in the magnetic circuit. This effect is called hysteresis. The loss of energy shows itself as heat in the specimen, and the engineer has to be aware of the problem and design his circuits accordingly. It can be shown that the magnitude of the losses generated is proportional to the area of the loops. The curves are known as *hysteresis loops*. An example of a magnetic material which is continually being subjected to reversals of current is a transformer core, and the material must be carefully chosen to avoid excessive heat.

Certain materials, for example a hard steel, retain a large proportion of the flux density achieved at saturation. On the contrary, only a small retained flux density occurs when a core of soft iron is used. Thus, not only do the values at corresponding points on the loop vary depending on the material, but so does the overall shape of the hysteresis loop. Typical loops for the examples given above are shown in Fig. 20.

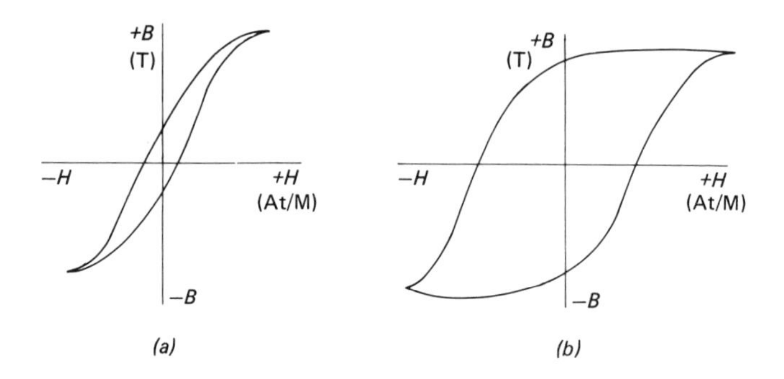

Fig. 20. *Hysteresis loops for:* (a) *a soft magnetic material; and* (b) *a hard magnetic material*

Very hard silicon steel will generate an even narrower loop and, for particular applications, ferrite materials have been developed having a loop which is virtually rectangular (*see* Fig. 21). This will display the property of almost instant saturation for very small values of magnetising current, and is particularly useful in the computer industry where two distinct states "magnetised" or "not magnetised" are needed.

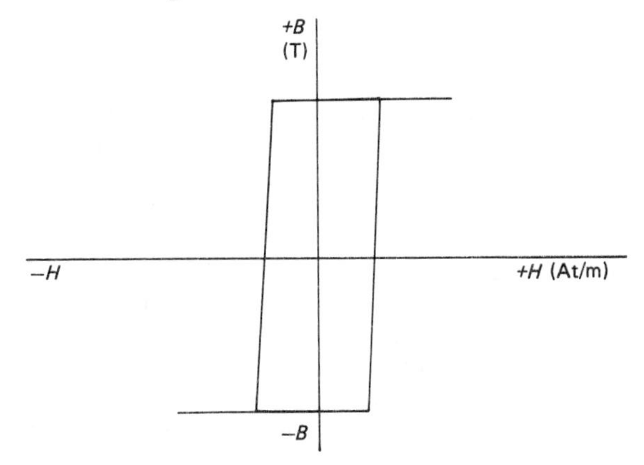

Fig. 21. *Hysteresis loop for a ferrite*

SELF-ASSESSMENT QUESTIONS

1. A current of 3.5 A flows in a coil of 400 turns. Calculate the value of the m.m.f.

2. The value of magnetising force in a coil of length 20 cm is 8×10^3 At/m. If the number of turns on the coil is 1,000, calculate the value of current required.

3. A lifting magnet is required to produce a flux density of 1.75T. If the cross-sectional area of the magnet is 0.8 m^2, calculate the value of magnetic flux required.

4. The effect of inserting a steel core inside an air-cored coil will be:
   (*a*) to increase the value of the coil reluctance;
   (*b*) to increase the value of the magnetic flux;
   (*c*) to decrease the value of the permeability;
   (*d*) to demagnetise the coil.

5. An electromagnet has a flux density of 1.5 T and a relative permeability of 1,700. Calculate the value of the magnetising force.

6. The effect of hysteresis in an iron-cored coil is:
   (*a*) to produce heat in the iron core;
   (*b*) to reduce the flux to zero;
   (*c*) to reverse the polarity of magnetisation;
   (*d*) to make it more difficult to magnetise the core.

7. An iron ring has a mean diameter of 15 cm and cross-sectional area of 4 cm^2. The magnetisation curve for the iron is as follows:

B (T)	0.45	0.8	0.98	1.08	1.15	1.2
H (At/m)	97	205	310	395	495	600

Calculate the value of current required to produce a flux of 4×10^{-4} Wb in the steel if the ring has a uniformly wound coil of 200 turns.

8. Calculate the reluctance of a closed magnetic circuit having a cross-sectional area of 2.2 cm^2 and an effective magnetic length of 110 cm. The relative permeability of the core is 1,250 and a current of 2 A is passed through a coil of 450 turns wound on the core. Also calculate the value of magnetic flux.

9. Draw typical hysteresis loops for cast iron, cast steel and motor armature laminations, listing the characteristics which make these materials different magnetically.

Indicate which material will have the greatest power loss when subjected to alternating magnetisation.

CHAPTER THREE

Capacitors and Capacitance

CHAPTER OBJECTIVES

After studying this chapter you should be able to:
* appreciate the basic principles and terminology of electrostatic fields;
* appreciate the application of electrostatic fields to capacitors;
* determine the electric field strength, electric flux density and capacitance for given conditions and dimensions;
* solve problems involving multiple capacitor combinations;
* determine the energy stored in given capacitors;
* appreciate the practical make-up and properties of a range of modern types of capacitor.

All of us, at some time, have picked up a comb and, after combing our hair, noticed that the hair tends to follow the comb as it is moved away from the head. In some way it seems to be attracted to the comb. In fact it is, the reason for this being that the comb and the hair have become electrically charged and, since certain charges attract each other, this pulling effect occurs. Let us look more closely at the reasons for this, and why materials are said to be charged.

ELECTRIC FIELDS AND ELECTRIC CHARGES

Electric Charge

Try a simple experiment. Rub an ordinary plastic pen quite vigorously on your sleeve for a few seconds, then put it amongst a few small scraps of paper. You will immediately notice that the pieces of paper are quite strongly attracted to the pen. Now knock off the pieces of paper and firmly grip the part of the pen where the pieces of paper used to be. If you bring the pen close to the pieces of paper now, you will find that they are no longer attracted.

There must be a logical explanation for these results. What has happened to cause this effect? The physical action of rubbing the pen on the sleeve has displaced some of the electrons from the atoms of the sleeve material, and they are acquired by the plastic pen casing. The casing now has an excess of electrons, and the material is said to be negatively charged. Similarly, the area of the sleeve which has lost the electrons will have an excess of positive charges, and is said to be positively charged.

It is possible, then, to produce a displacement of electrons to or from a body (which may be conducting or non-conducting), and to build up a positive or negative charge on it.

Going back to our original example, when the negatively charged pen was gripped, the excess electrons flowed through the person to earth and the pen returned to its original state of neutral charge. There was no longer any attraction for the pieces of paper and attempts to pick them up failed.

Electric Fields

Let us do another simple experiment. Take two plastic pen tops and tie pieces of cotton around the pocket clip. Now, holding only the end, vigorously rub each in turn on your sleeve. Using the cotton, suspend both tops and carefully move them close to each other. Once more an interesting effect is noticed. As the pen tops are brought together they "avoid" each other. There is no contact between the bodies, but one obviously effects the other. In this case it is a force of repulsion, and it must be deduced that there is an influence of one charged body on the other. This influence is felt some distance away from the body and, therefore, a type of field similar to the magnetic field must exist. In this case the field is generated by the displacement of electric charges; hence it is called an *electric field.*

As in the case of the magnetic field, the forces produced act in all directions from the charged body, but to help us have some

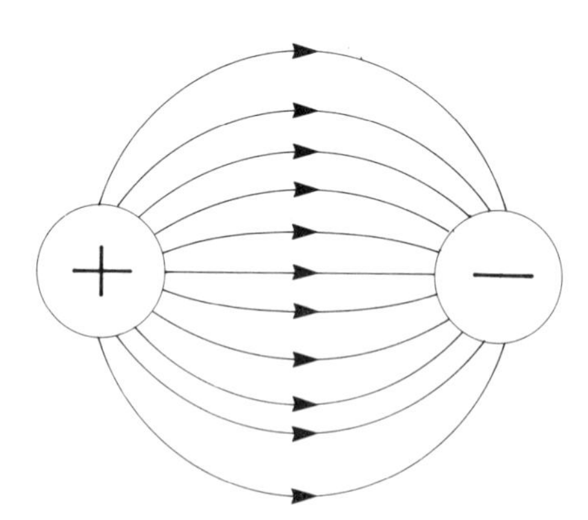

Fig. 22. *The electrostatic field between charges*

means of showing this in a diagram, we draw individual lines of force. The lines are always shown as starting from positive charges and terminating on negatively charged bodies (*see* Fig. 22). Between similarly charged bodies, such as the two pen tops, the force is one of repulsion, whereas between two bodies holding unlike charges the force will be attractive.

THE CAPACITOR

All our previous study has considered that, to be effective, an electrical circuit must be continuously connected by physical conductors. Let us now consider a simple series circuit, such as the one shown in Fig. 23. The ends of the conductors at A and B

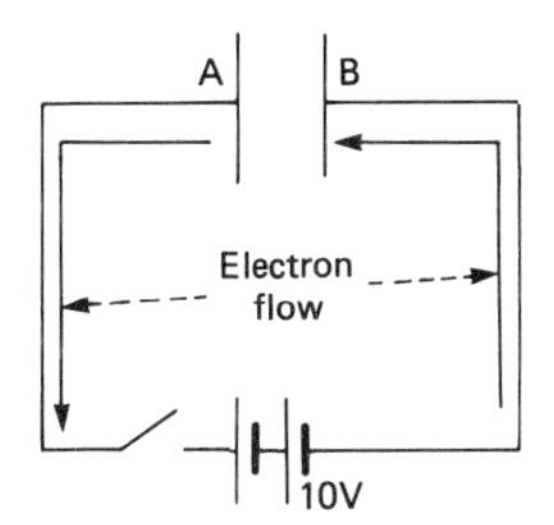

Fig. 23. *The action of charging a capacitor*

terminate in two flat conducting plates, separated by a small air gap which are together known as a capacitor. It would seem reasonable to consider that nothing will occur when the switch is closed. However, something does happen and it is this that we need to study in more detail.

Charging Current

When the switch closes, the positive terminal of the battery is 5 V more positive than the plate A. However, we know that when a potential difference occurs across a conductor a current will flow, and this situation is no exception.

Negative electrons are attracted towards the positive terminal of the battery, and flow in the direction shown. Electrons leaving plate A cause it to become more positive, thus reducing the potential difference between it and the positive terminal of the battery. As the potential difference is reduced, so the flow of electrons slows. However, this flow will continue at an ever-decreasing rate until there is no potential difference, i.e. A is at 5 V. During the same time the negative terminal of the battery has been feeding out electrons causing the plate B to become more negative. When B attains the same potential as the negative terminal of the battery, the current ceases and the total potential across AB is equal to the battery voltage. The negative charge gained by B is exactly equal to the negative charge lost by plate A.

Plates A and B now have a potential across them and an electric field exists between them. It establishes a field pattern as shown in Fig. 24.

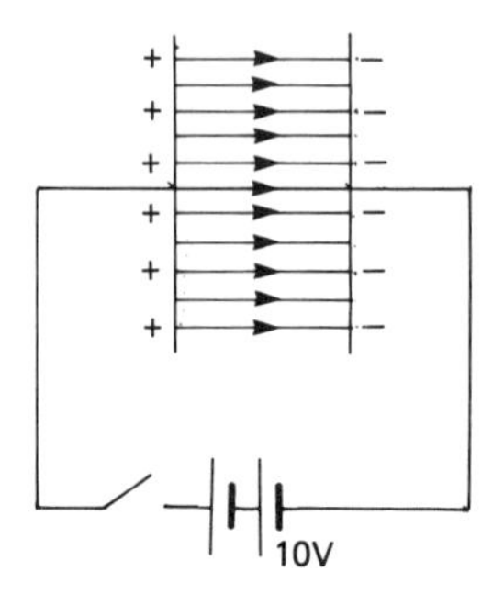

Fig. 24. *The electrostatic field between capacitor plates*

Energy has been used in setting up this charge pattern between the conducting plates. When built in this fashion, the plates are known as a capacitor. The energy is stored in the electric field, and can be reclaimed by discharging the capacitor, as will be discussed later.

The charging current is not constant. It is high at first and reduces as the magnitude of the charges on the plates increases. A typical charging current is displayed graphically in Fig. 25 and, on the same graph, the rise of potential difference across the plates is also plotted. The time taken to charge the capacitor can

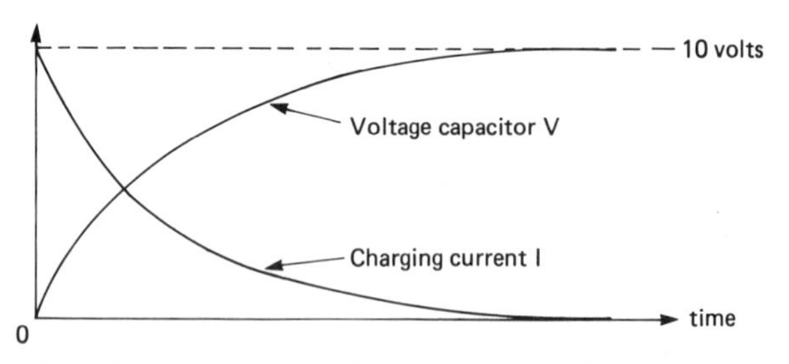

Fig. 25. *Charging current and capacitor voltage curves*

vary from fractions of a second to several minutes, but it is not usually more than a few seconds, and depends upon the circuit conditions. The relationship between the amount of charge held and the p.d. across the plates is a measure of the capacitance. The unit of capacitance is the *farad* (F) and you will remember from Chapter 1 that this is too large a unit for practical use.

Electric Flux Density (D)

The charge set up on the plates is measured as Q C, and the electric field which is set up as a result of this charge is known as the electric flux. It is given the Greek symbol Ψ (psi). The electric flux Ψ generated from a charge of Q C is measured as Q C. As in the case of the magnetic field, the total flux does not really give a good indication of the overall strength, and it is the amount of electric flux passing through a given area which is important. This is known as the electric *flux density* and is given the symbol $\boldsymbol{D}$. $\boldsymbol{D}$ is therefore the flux per unit area:

$$\text{and } \boldsymbol{D} = \frac{\text{total flux}}{\text{area of plates}} = \frac{Q}{A}\ \text{C/m}^2.$$

Electric Field Strength (E)

When two metal plates, set some distance apart, are charged up, a potential difference will exist between them. A force will thus be generated between them, which is described as the electric force or electric field strength. This is given the symbol $\boldsymbol{E}$. The size of the force will obviously depend upon the potential difference across the plates, and also on the distance between them. The influence of one plate on the other will not be as great if a large distance separates them, as it would be if only a small gap existed. This physical make-up is very important in the capacitor.

The force, or electric field strength, can be deduced by using the following expression.

$$\text{Electric field strength } \boldsymbol{E} = \frac{\text{Potential difference}}{\text{distance between plates}} = \frac{V}{d}\ \text{V/metre}$$

Potential Gradient

This quantity, V/d volts per metre, is often said to be the *potential gradient* between the plates. Potential gradient is very important when the breakdown of insulation properties is being considered. An insulator may well withstand a considerable voltage across it, when the distance between the conducting surfaces which it is protecting is quite large. Should the potential gradient increase considerably, either because the voltage increases or the distance between the surfaces reduces, breakdown of the insulation material may occur.

The thickness of the insulation and the position of the conductors relative to other conducting surfaces are therefore very important to both the electrical power engineers and the electronics design engineers. Although the electronics designer may deal only with 3 or 4 V, the distances between conducting surfaces are often very small indeed, and damaging potential gradients may result. A typical example is in the very small gaps of one or two thousands of an inch inside integrated circuits.

In power applications, voltages of almost half a megavolt are used. The gaps in these cases have to be very large. Overhead power lines have considerable capacitances between their cables and this fact must be taken into account in the design of them.

The Prime Mover and the End Product in the Electric Circuit

In a previous discussion concerning magnetic circuits, the idea that the magnetising force was the prime mover and that the magnetic flux density the end product, was introduced. A very similar parallel can be drawn when looking at electric fields. The electric field strength or potential gradient, is the prime mover. It is the application of a potential difference across the plates which produces a charge on them, and develops an electric field between them. The end product is this electric field, the strength of which is measured in terms of the overall flux density.

The ratio comparing the resulting flux density with the electric field strength applied, is a measure of how effective is the capacitor that we are producing.

Dielectrics

The insulating material between the two conducting plates of a capacitor is called a *dielectric*. The electric field is established within this dielectric, and the eventual flux density achieved will vary quite considerably, depending upon the particular material used.

The storage of energy in a capacitor is achieved by deformation of the electron orbits associated with the individual atoms of the dielectric material. It is only possible to give a general description of this storage effect in this chapter, because the study of electron orbits is very complex.

Figure 26(*a*) shows the nucleus of an atom and a single

electron, following a circular path. Let us assume that this is the pattern inside the dielectric between the plates of the uncharged capacitor.

The capacitor is next charged. One plate becomes positively charged, the other negatively charged. The orbiting electron also has a charge, which we know is negative. Therefore, it is subjected to forces due to the charges on the conducting plates. It is repelled by the negative plate, but attracted by the positive plate. The electron orbit now follows the pattern shown in Fig. 26(*b*). External energy is required to produce this deformation,

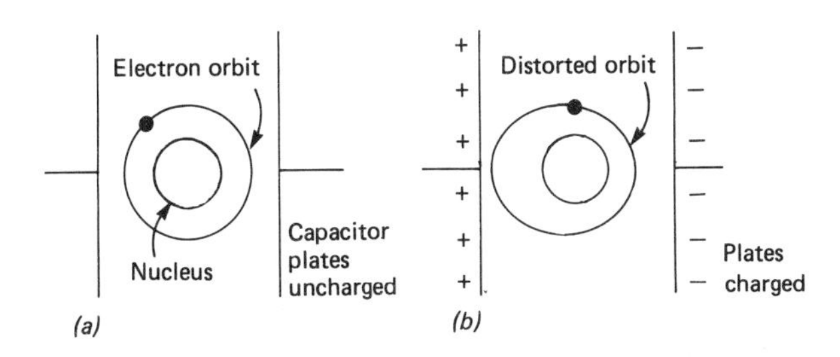

Fig. 26. *The principle of energy storage in a capacitor dielectric showing electron orbits with:* (a) *plates charged; and* (b) *plates uncharged*

but it can be reclaimed when the capacitor is discharged. The orbit returns to the original condition, and the dielectric resumes its original state.

Permittivity

This energy storage ability, varies for different dielectric materials, and will control the overall capacitance which a capacitor can have. The value of the capacitance can be made to vary if slabs of various dielectrics are placed between the two metal plates. The ratio of "success to effort", is therefore a ready means of expressing the efficiency of a particular dielectric material. When all other factors remain the same, the properties of the dielectric govern the overall capacitance.

This efficiency ratio is known as the permittivity of the dielectric material and it is defined as:

$$\varepsilon = \frac{D}{E} = \frac{\text{Electric flux density}}{\text{Electric force}}.$$

The unit of permittivity is the farad/metre.

The Relationship between Charge, Potential Difference and Capacitance

As the potential difference across the plates increases so does the charge on a given capacitor. They are directly proportional to each other. A doubling of the voltage doubles the charge. The relationship can best be described in the expression given here:

$$Q = VC$$

where: Q is the charge in coulombs;
V is the potential difference across the capacitor plates in volts; and
C is the capacitance in farads.

The unit of capacitance, the farad, can be defined as the capacitance of a capacitor that, when given a charge of 1 coulomb, produces a potential difference of 1 volt between its plates.

How the Capacitance is Governed by the Physical Dimensions of the Capacitor

Let us look again at the equation given, and expand it by substituting the definitions for D and E which have already been stated.

We have seen that $$\varepsilon = \frac{D}{E}.$$

Using $$D = \frac{Q}{A}$$

and $$E = \frac{V}{d}$$

we see that $$\varepsilon = \frac{Qd}{AV},$$

but $Q = VC$ $$\therefore \quad C = \frac{Q}{V}$$

Substituting for $\frac{Q}{V}$ $$\varepsilon = \frac{Cd}{A}$$

$$\therefore \quad C = \frac{\varepsilon A}{d}$$

Where C is the capacitance in farads; A is the area of the plates in square metres; d is the distance between the plates in metres; and ε is the permittivity of the dielectric in farads/metre.

We can now carefully examine the physical factors which will influence the capacitance of a given capacitor.

The capacitance is directly proportional to the area of the plates. The bigger the area, the greater the capacity.

The capacitance is inversely proportional to the distance between the plates. When the gap between the plates is made larger, the capacitance will be reduced.

The choice of dielectric material will influence the capacitance. The capacitance is directly proportional to the permittivity. This means that dielectrics with a high value of ε are better from a design point of view.

Relative Permittivity

Let us go back to the basic expression $\varepsilon = D/E$. You will remember that this is the ratio "success to effort".

Now let us assume that we have set up two metal plates, connected to a circuit, as the two plates of a capacitor. By putting a series of different dielectric materials between them, we can find the permittivity of those materials, and how effective they are for use as dielectrics.

When we have tried several materials, we can calculate the ratio ***D/E***, i.e. the permittivity ε for each of the materials and arrange the values thus obtained in order.

As in the case of the magnetic circuit, the actual figures are extremely clumsy, a typical example being 0·000000000032. Such figures are inconvenient, and exactly the same kind of technique is used as was used for the permeability values in the case of magnetic fields.

The principle is quite simple. A vacuum is set up between the plates, ***D*** and ***E*** are measured and the value of ε is calculated. This value is taken as the standard and is given the special symbol ε_0, and is known as permittivity of free space. Its value is $8{\cdot}85 \times 10^{-12}$ farad/metre and it never varies.

We have found the permittivity of free space, ε_0, and by calculating the value of the ratio $\varepsilon/\varepsilon_0$ for each material, we can obtain a quantity which is specific to each dielectric and which is known as the relative permittivity (symbol ε_r). Thus, the absolute permittivities of all materials can be expressed as the product of the permittivity of free space, ε_0 (an awkward quantity), and the relative permittivity, ε_r (which is a relatively simple quantity).

It should be noted that the permittivity of air is very little different from that of a vacuum, and that this difference is usually ignored. The relative permittivity of air is therefore usually assumed to be unity for all practical purposes.

Example 1

Calculate the capacitance of a capacitor which has two plates each of cross sectional area 150 cm², when a dielectric 0·5 mm thick is placed between them. ε_r for the material is 12.

$$\text{From } C = \frac{\varepsilon_0 \varepsilon_r A}{d} \text{ farads;}$$

where ε_0 = permittivity of free space (F/m);
ε_r = relative permittivity; s (m²);
A = active area of plates (m²);
d = distance between plates (m).

$$C = \frac{8.85 \times 10^{-12} \times 12 \times 150 \times 10^{-4}}{0.5 \times 10^{-3}} \text{ F.}$$

$$= \frac{8.85 \times 12 \times 15}{0.5} \text{ pF}$$

$$= 3{,}186 \text{ pF} \quad \text{or} \quad 3.186 \text{ nF.}$$

It is important to note again that the farad is a very large unit and microfarads (10^{-6}), nanofarads (10^{-9}) or picofarads (10^{-12}) are very commonly used when capacitances are being considered.

Multi-plate Capacitors

It has already been explained, that to get a high capacitance, the area of the conducting plates should be as large as possible. This now poses a problem. The plates become physically large to handle, and some alternative method of increasing the effective area has to be found.

A simple but logical method is employed. The plates are reduced in size by chopping them up into smaller sections, placing them side by side interleaved with dielectric, and then by connecting alternate plates together, the total area is used.

A common example of this is the air-spaced capacitor often used as a tuning capacitor in radio engineering. Fig. 27 shows the basic construction.

Since a capacitor needs conducting plates and insulating dielectric, it must be noted that although there are 11 plates shown, there are only 10 effective capacitors. The expression used previously has to be modified to take account of this.

Hence for a multi-plate capacitor:

$$C = \frac{\varepsilon_0 \varepsilon_r A (n-1)}{d} \text{ F}$$

where n is the number of plates.

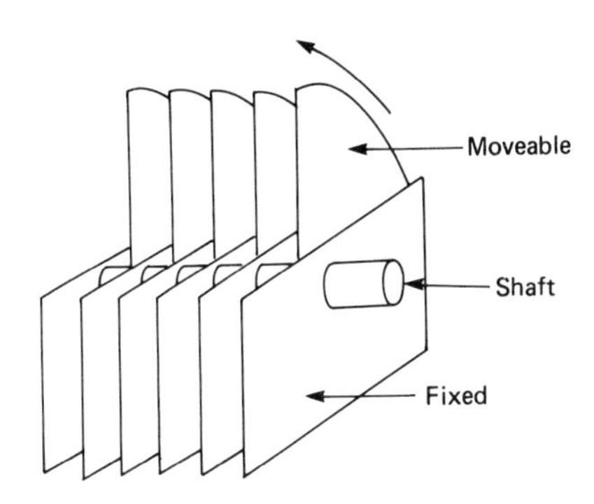

Fig. 27. *The air dielectric variable capacitor*

Example 2

A multi-plate capacitor has 11 plates, each of 100 cm². The dielectric is a material having $\varepsilon_r = 5$, and the distance separating each pair of plates is 2 mm. Calculate the capacitance, and the charge which will appear on it when a d.c. potential of 20 V is applied across the terminals.

$$\text{From } C = \frac{\varepsilon_0 \varepsilon_r A (n-1)}{d}$$

$$= \frac{8.85 \times 10^{-12} \times 5 \times 100 \times 10^{-4} \times 10}{2 \times 10^{-3}} \text{ F}$$

$$= \frac{8.85 \times 5 \times 100}{2} \text{ pF}$$

$$= 2212.5 \text{ pF or } 2.2125 \text{ nF.}$$

The charge on it can be obtained from: $Q = VC$, where: Q = charge (C); V = applied voltage (V); C = capacitance (F). Therefore $Q = 20 \times 2{\cdot}2125 \times 10^{-9}$ C = 44·25 nC

SERIES AND PARALLEL COMBINATIONS OF CAPACITORS

The Effect of Connecting Capacitors in Series

We know that in any series circuit, the sum of the individuals voltages distributed around the circuit must equal the voltage applied (*see* Fig. 28).

$$\text{Hence, } V_T = V_1 + V_2 + V_3$$

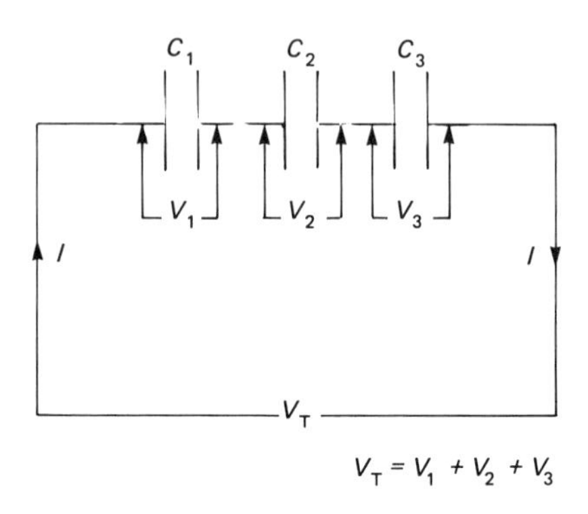

Fig. 28. *Voltage distribution for capacitors in series*

Now the voltage across each capacitor will depend on the charge held by it, and on the actual value of the capacitance,

i.e. $$V_1 = \frac{Q_1}{C_1},\ V_2 = \frac{Q_2}{C_2}, \text{ etc.},$$

and we can write $$\frac{Q_T}{C_T} = \frac{Q_1}{C_1} + \frac{Q_2}{C_2} + \frac{Q_3}{C_3}$$

We know that we cannot stockpile electrons anywhere in the conductors, so that whatever charge accumulates on C_1 must be on C_2, C_3, etc. and $Q_T = Q_1 = Q_2 = Q_3$

$$\therefore \quad \frac{Q}{C_T} = Q\left(\frac{1}{C_1} + \frac{1}{C_2} + \frac{1}{C_3}\right)$$

$$\therefore \quad \frac{1}{C_T} = \frac{1}{C_1} + \frac{1}{C_2} + \frac{1}{C_3}$$

This means that when capacitors are connected in series, the total capacitance C_T must be calculated by adding their reciprocals and inverting.

It should be noticed, that this is an expression which is very similar to one which we have met previously for calculating the total effective resistance of resistors connected *in parallel*, whereas this is for capacitors connected *in series*.

Example 3
Find the total effective capacitance when two capacitors are connected in series, if they have capacitance of $20\,\mu\text{F}$ and $30\,\mu\text{F}$ respectively.

From
$$\frac{1}{C_T} = \frac{1}{C_1} + \frac{1}{C_2}$$

$$\frac{1}{C_T} = \frac{1}{20} + \frac{1}{30}$$

$$= \frac{3+2}{60}$$

$$\therefore \quad C_T = \frac{60}{5} = 12\,\mu\text{F}.$$

Capacitors Connected in Parallel
In just the same way as resistors are often connected in parallel, then capacitors too can be connected in this way for various purposes (*see* Fig. 29).

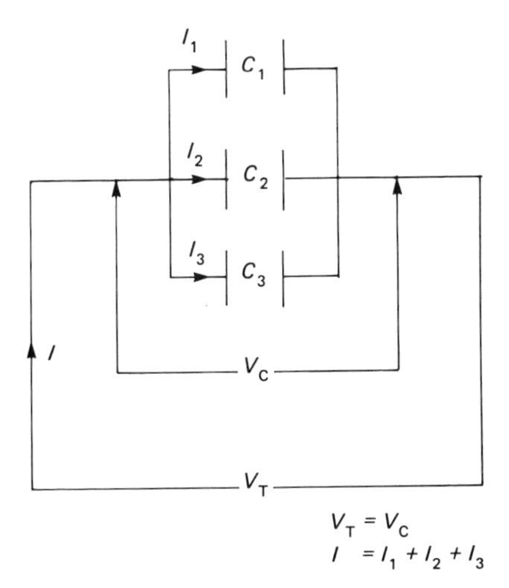

Fig. 29. *Current distribution for capacitors in parallel*

When any components are connected in parallel, the voltage across all of the individual parallel branches must be the same. This, then, is our starting point. The charge on each capacitor for a constant value of V will depend on the size of the capacitor.

$$Q_T = Q_1 + Q_2 + Q_3$$

But from $Q = VC$ $\quad C_TV = C_1V + C_2V + C_3V$

and $\quad C_TV = V(C_1 + C_2 + C_3),$

but we know V is constant.
Hence $\quad C_T = C_1 + C_2 + C_3,$

i.e. to find the total capacitance of several capacitors connected in parallel, we simply sum their individual capacitances.

It must be appreciated that in the expressions and examples given, only two or three capacitors are used. The rules hold good for any number and are applied in just the same manner.

Series-parallel Combinations

It may be necessary to connect a group of capacitors in a combination of series and parallel groupings, to achieve a particular value.

Example 4

Calculate the total capacitance value of the circuit configuration given in Fig. 30

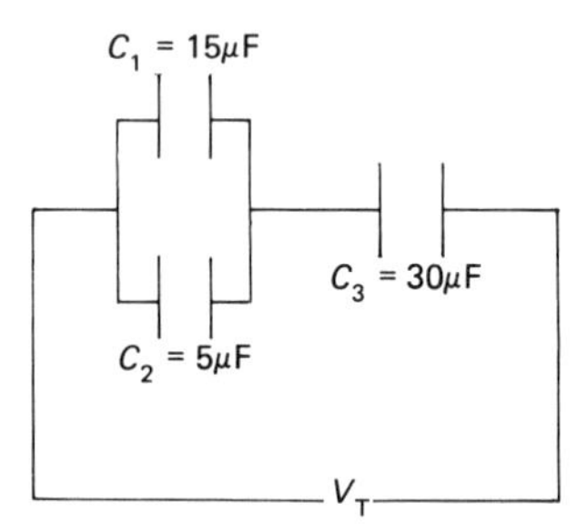

Fig. 30. *Calculation of the total value of capacitance*

As in a similar case using resistors, the capacitance of the parallel grouping must be calculated first, then added to the series component.

Hence from

$$C_{\text{parallel}} = C_1 + C_2$$
$$= 15 + 5$$
$$= 20\,\mu\text{F}.$$

The combined capacitance of the parallel group is therefore $20\,\mu$F and this must now be combined with the $30\,\mu$F of C_3.

From

$$\frac{1}{C_\text{T}} = \frac{1}{C_{\text{parallel}}} + \frac{1}{C_3}$$
$$= \frac{1}{20} + \frac{1}{30}$$
$$= \frac{3+2}{60}$$
$$= \frac{5}{60}$$
$$\therefore \quad C_\text{T} = \frac{60}{5} = 12\,\mu\text{F}.$$

Dielectric Loss

We have already discussed the idea that when a capacitor is in the charged state, the electron orbits are subjected to a deformation which is the main form of energy storing. When the capacitor is discharged, the energy is given back to the circuit and the electron orbit returns to normal.

Not quite all of the energy is returned, and a small amount is dissipated as heat in the dielectric. This is known as *dielectric loss*. There may well be a conflict between dielectrics which have a high permittivity, and those which have a low loss. The manufacturer has to decide, when choosing his materials, which priority he wishes to give.

Safe Working Voltages

Previously, we have mentioned that when the potential gradient becomes sufficiently high, even a good insulating material may be subject to breakdown by a spark current. This will often damage the physical structure of the insulator such that it no longer retains its insulating properties.

A capacitor may often have a high potential difference across its plates, and because it has a small distance between them (in order to get high values of capacitance), the potential gradient may become very large. For this reason, the manufacturer will quote a maximum voltage which can be safely applied, without breaking down the insulation. These are known as safe working voltages and are usually stamped on the capacitor itself.

Breakdown potential gradients are quoted in kV/mm, but these are the figures which are used by the manufacturer when considering the choice of dielectric material, and safe working voltage is of more use to the user.

The Amount of Energy Stored in the Capacitor

Assume that the charge on a capacitor is increased in small steps. The voltage will increase accordingly. The area under the graph shown in Fig. 31 at any time is equal to the energy stored.

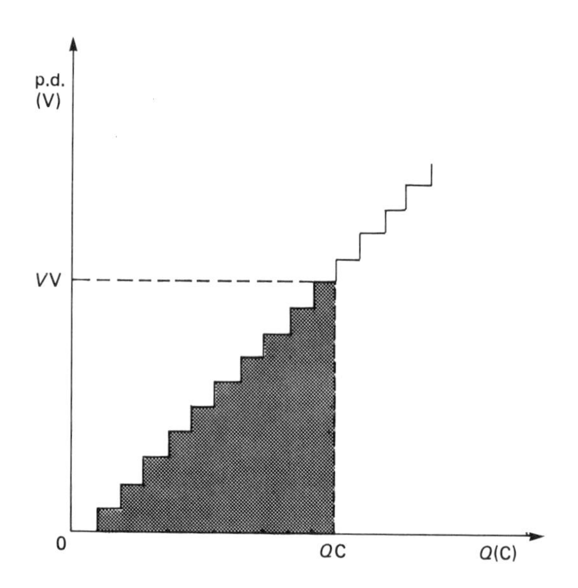

Fig. 31. *Relationship between p.d. and charge*

Therefore, when a p.d. of V volts is established, the energy stored is equal to the shaded portion and:

$$E = \frac{1}{2} Q \times V$$

but $Q = CV$

$$\therefore \quad E = \frac{1}{2} CV \times V$$

$$= \frac{1}{2} CV^2 \text{ J.}$$

where C = capacitance in farads
and V = applied voltage in volts.

TYPES OF CAPACITOR

We have looked at various aspects which the manufacturer of capacitors has to consider when the capacitor is designed. These are summarised as follows:

1. large plate area for high capacitance;
2. small plate area to reduce physical size;
3. small plate separation for high capacitance;
4. as large a plate separation as possible for high breakdown voltage;
5. high permittivity dielectric for high capacitance;
6. as low a dielectric loss as possible.

A conflict often exists as, for example, between 1 and 2, and a manufacturer simply has to make the product which he thinks will meet the market requirements.

Air Dielectric Capacitors

A further factor to consider is the range of frequencies which the capacitor is likely to meet, e.g. zero in a d.c. circuit or very high in the case of radio transmitters. Where this is the case, a solid dielectric would almost certainly produce such large losses that the amount of heat generated could not be tolerated, and air dielectric capacitors would probably be used. Air has a relative permittivity of unity, and the capacitors would therefore, have a low overall capacitance, ranging from about 10–20 pF. These capacitors are often known as "trimmers". Air capacitors have capacitances up to about 1,000 or 2,000 pF.

Waxed Paper/Foil Type

One very convenient method of increasing the effective plate area is to make the plates thin enough to be flexible, sandwich a wax impregnated paper between them and roll them up. This has been common practice in manufacture for a considerable period and many capacitors in use are manufactured in this way. Figure 32 shows the construction of this type of capacitor.

A range of capacitances from 50 pF up to around 16 μF is available from waxed paper types.

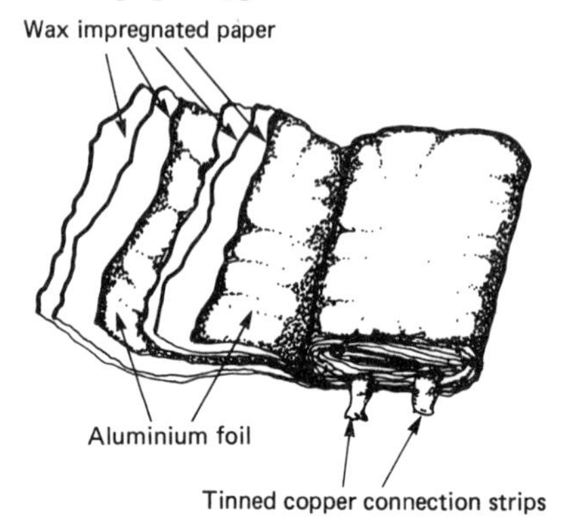

Fig. 32. *Construction of the wax paper/foil type of capacitor*

Polyester Type

The waxed paper has basic disadvantages. It is hygroscopic and has relatively low safe working voltages. This has given rise to the development of the polyester/foil type. The structure is very similar to that of the waxed paper type, being a rolled assembly. Polyester is used as the dielectric and is comparatively loss free. It has a higher permittivity than waxed paper, which means that high values of capacitance can be achieved with the same physical dimensions. Polyester/foil capacitors can be obtained with capacitances up to several microfarads.

Mica Types

A dielectric of mica can provide useful properties (*see* Fig. 33). It has high dielectric strength and also a high permittivity.

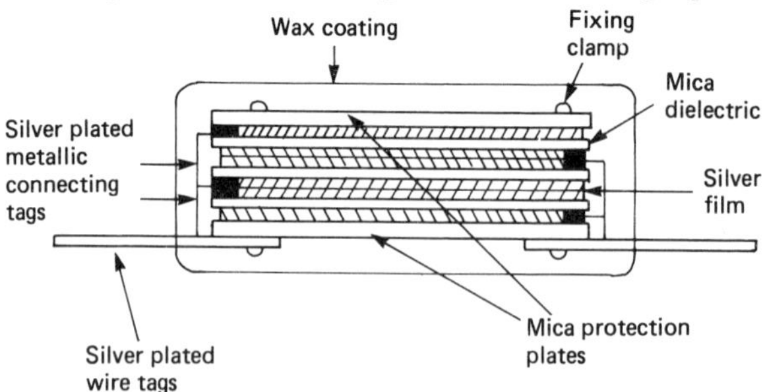

Fig. 33. *An example of the mica type of capacitor*

However, because it is too rigid a material to be rolled the higher permittivity is effectively cancelled out because it has to be made physically larger to produce high capacitances. It is usually made with a parallel plate type of assembly with foil conducting plates. Alternatively films of silver can be deposited directly on to the mica sheets by an evaporation process. Close tolerance capacitors can be produced in this way giving good all round stability, usually with capacitances between 10 and 20 nF.

Ceramic Types

Certain modern ceramics have extremely high permittivity values and this has led to a group of relatively small high value capacitors. A disc or tubular construction is commonly used (*see* Fig. 34). Typically capacitances of 1–500 nF are available from this type of capacitor.

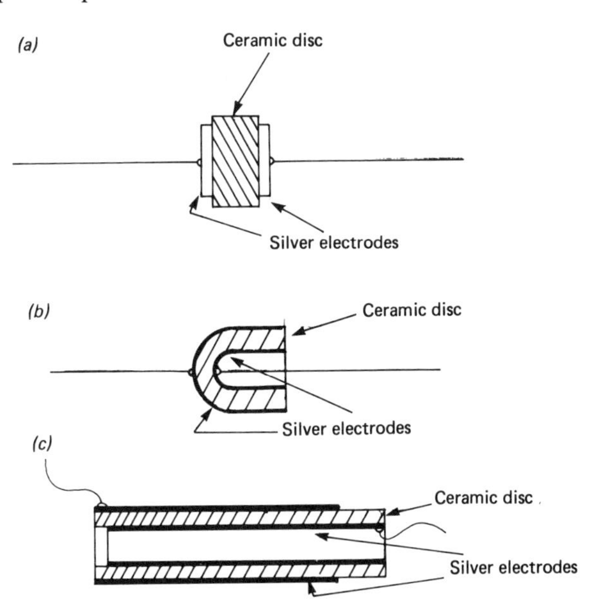

Fig. 34. *Examples of the ceramic type of capacitor:* (a) *disc type;* (b) *hat (or cup) type; and* (c) *tube type*

Electrolytic Capacitors

It has long been known that the oxide layer which forms on the surface of conducting metals, produces a good insulating surface. The fact that this oxide layer is of extremely narrow dimension was recognised as being a possible method of producing a capacitor. The distance *d* between the plates is obviously very small, giving very high capacitance values. The oxide itself acts as the dielectric and is maintained by the applied voltage (*see* Fig. 35).

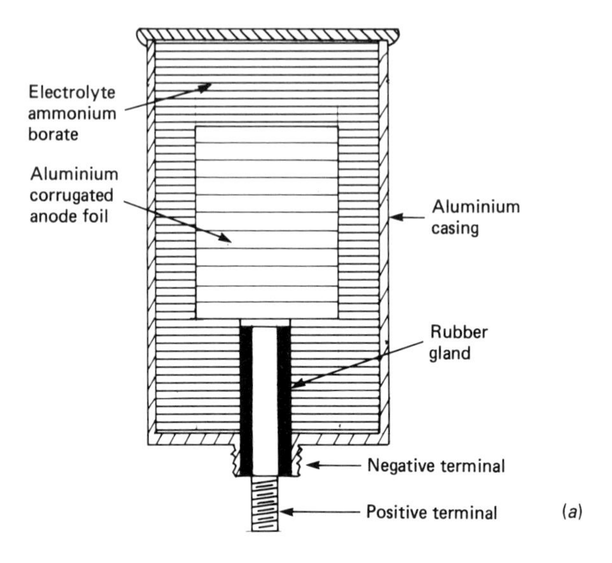

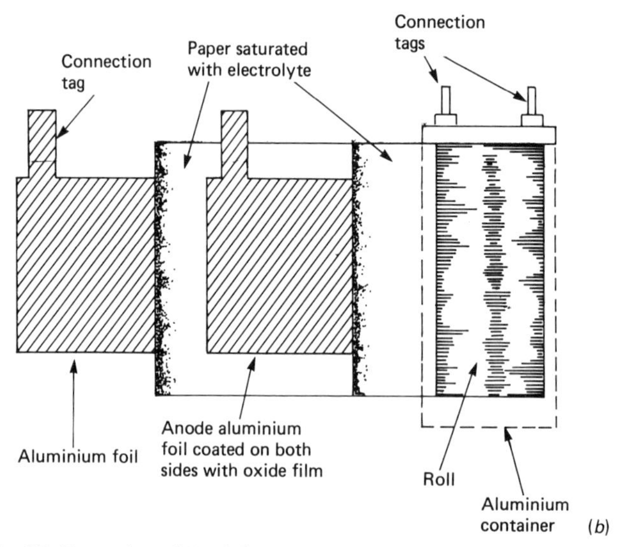

Fig. 35. *Examples of the:* (a) *wet; and* (b) *dry types of capacitor*

The layer is produced only when the applied voltage is connected the correct way around, giving rise to the major disadvantages of electrolytic types, that they are polarised, and should not be reverse connected.

Capacitance values range from about 1 μF upwards to many thousands of microfarads and are probably the major type of capacitor for all applications requiring high capacitances.

Leakage

Most capacitors have the disadvantage that when left in a charged state the tendency is for the charge to gradually leak away through the dielectric. The ability to withstand this is desirable in a capacitor. The electrolytic group of capacitors suffer the disadvantage that they are probably the most prone to this effect.

CAPACITORS IN SIMPLE CIRCUITS

It is often important to be able to calculate the voltages at various points in the circuit when groups of capacitors are connected in series, parallel or series-parallel groupings. As usual, the solution follows straightforward rules, but it is important to think of the principles we have considered in this section, before the rules are applied. The most important of these are:

(*a*) That in a series circuit the capacitors must all take up the same charge. We cannot bunch up electrons around the circuit, hence when we charge up the capacitors, assume that the capacitor is charged by a constant current I for a given time t. The charge in coulombs must then be $I \times t$. Hence the charge on each capacitor is the same. This can be summarised:

Series circuit—constant charge; potential difference varies.

(*b*) In a parallel circuit the voltage across each capacitor must be constant but the charging current through the branches will vary and hence the eventual charge will vary.

In summary: parallel circuits—constant voltage; charge varies.

Example

Calculate:

(*a*) the total capacitance;
(*b*) the voltage across each capacitor;
(*c*) the charge on each plate of C_1, C_2 and C_3;
(*d*) the energy stored in C_1;

for the circuit shown in Fig. 36.

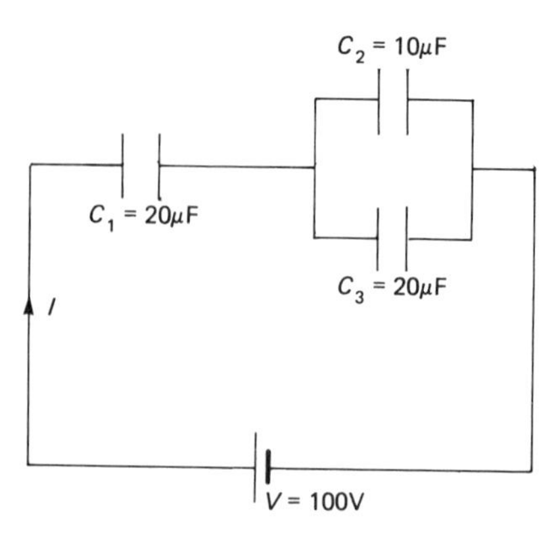

Fig. 36. *Circuit diagram for Example*

First, as before, calculate the combined capacitance of the parallel group

using
$$C_p = C_3 + C_2$$
$$= 20 + 10$$

therefore
$$C_p = 30\,\mu\text{F}$$

The second step is to combine the $30\,\mu$ F calculated above with the $20\,\mu$ F of C_1. Here we use the expression:

$$\frac{1}{C_T} = \frac{1}{C_1} + \frac{1}{C_p}$$
$$= \frac{1}{20} + \frac{1}{30}$$
$$= \frac{3+2}{60} = \frac{5}{60}$$

therefore
$$C_T = \frac{60}{5} = 12\,\mu\text{F}$$

and
$$Q_T = V \times C_T$$
$$= 100 \times 12$$
$$= 1200\,\mu\text{C}$$

$1200\,\mu$C must therefore exist on each plate of C_1 since all components in series have this common charge. Hence we can now look at the individual components and calculate the p.d. across each capacitor.

From
$$Q = VC$$
$$V = \frac{Q}{C}$$
$$V_1 = \frac{Q}{C_1}$$
$$= \frac{1{,}200 \times 10^{-6}}{20 \times 10^{-6}}$$
$$V_1 = 60\,\text{V}$$

Similarly, the total charge on each side of the parallel group will be 1,200 μC and

$$V_p = \frac{Q}{C_p}$$
$$= \frac{1{,}200 \times 10^{-6}}{30 \times 10^{-6}}$$
$$V_p = 40\,\text{V}.$$

and V_p and V_1 add up to 100 V, the applied voltage. The capacitors C_2 and C_3 will each have a p.d. of 40 volts across them since they are in parallel.

We can now determine the charge on each. We already know that 1,200 μC is the charge on each side of C_p, but this charge will be distributed between the elements of the parallel group and,

$$Q_1 = V_1 C_1$$
$$= 40 \times 10 \times 10^{-6}$$
$$= 400\,\mu\text{C}$$
$$Q_2 = V_2 C_2$$
$$= 40 \times 20 \times 10^{-6}$$
$$= 800\,\mu\text{C}$$

$400 + 800 = 1{,}200\,\mu$C which is the total charge on each side.

Lastly, we can determine the energy stored.

Energy $= \frac{1}{2}CV^2$ joules for each capacitor:

$$\text{for } C_1 = \frac{20 \times 10^{-6} \times 60 \times 60}{2} = 36\,\text{mJ};$$

$$\text{for } C_2 = \frac{20 \times 10^{-6} \times 40 \times 40}{2} = 16\,\text{mJ};$$

$$\text{for } C_3 = \frac{20 \times 10^{-6} \times 40 \times 40}{2} = 8\,\text{mJ};$$

Once again the example has been solved by using a logical approach and by remembering the basic conditions. You should now attempt the questions given as practice in dealing with series and parallel combinations of capacitors.

SELF-ASSESSMENT QUESTIONS

1. When two bodies of like charge are brought close to each other:
   (*a*) there will be no reaction;
   (*b*) the bodies will attract each other;
   (*c*) the bodies will discharge;
   (*d*) the bodies will repel each other.
2. The dielectric placed between the two plates of a capacitor:
   (*a*) reduces the charge which it will hold;
   (*b*) increases the capacitance;
   (*c*) decreases the current which will be needed to charge it;
   (*d*) changes the material from which the plates need be made.
3. A capacitor is charged from a 100 V supply and acquires a stored energy of 1 J. Calculate its capacitance and the value of the charge established on each plate of the capacitor.
4. A capacitor comprises two parallel plates each 10 cm square and the plates are spaced 3 mm apart. If the charge on each plate is 0.6×10^{-6} C, calculate:
   (*a*) the electric field strength between the plates;
   (*b*) the p.d. between the plates.
5. A parallel plate capacitor has two square plates each of side 20 cm which have a dielectric between them of relative permittively 5. The distance between the plates is 3 mm and a p.d. of 500 V is maintained across the plates. Calculate:

(*a*) the capacitance of the capacitor;
(*b*) the total charge stored on the plates;
(*c*) the electric field strength;
(*d*) the dielectric flux density.

6. Three capacitors, having capacitances of $10\,\mu F$, $20\,\mu F$ and $60\,\mu F$, are connected in series. Calculate the total capacitance. If this combination is connected to a 1,000 V d.c. supply, calculate the charge on the capacitors and the p.d. across each capacitor.

7. Show in a diagram how a large number of tin foil sheets may be interconnected to produce a capacitor of high value. A capacitor of this kind has 201 sheets of foil separated by mica sheets 0.3 mm thick, alternate plates being connected to opposite terminals. If the relative permittivity of the dielectric is 5, and the overall capacitance is $0.1\,\mu F$, calculate the effective area of each sheet.

8. Two parallel plate capacitors, A and B have the following dimensions:

(*a*) area of plates 1,200 cm^2, separation 4 mm;
(*b*) area of plates 3,000 cm^2, separation 2.5 mm.

The relative permittivity of the dielectric in A is 4.5 and in B is 4.0.

Calculate:

(*a*) the capacitance of each capacitor;
(*b*) the total capacitance if they are connected in series;
(*c*) the voltage across each capacitor if 750 V is applied across the capacitors when connected in series.

9. The current flowing in a capacitor of capacitance $20\,\mu F$, which is initially uncharged is kept constant at 1 mA for 3 seconds, then at 2.5 mA for the next 2 seconds and then during the next 4 seconds it is discharged. Calculate the average value of discharge current, if the capacitor is fully discharged. Calculate the potential difference across the capacitor at the end of each of these periods of time.

CHAPTER FOUR

Electromagnetic Induction

CHAPTER OBJECTIVES

After studying this chapter you should be able to:

* understand the basic principles of electromagnetic induction;
* appreciate how the basic operation of the motor, generator and transformer are linked to electromagnetic induction;
* quote the laws of Faraday and Lenz;
* calculate the force, voltage generated and transformer ratios when the physical parameters are known;
* define and explain the similarities and differences between self, and mutual, inductance.

In the previous section we considered the effects of the magnetic field, how it is set up and some of the rules governing how much flux is produced in given magnetic circuits. An important aspect of the magnetic field is that its effect is felt some distance away from the coil or conductor which is being used to produce it. The stronger the field is, the more pronounced is this effect. You will remember that we described this in terms of the magnetic flux density.

FARADAY'S DISCOVERIES

Many years ago, in about 1831, Michael Faraday, whose name all students of electrical engineering know, made a scientific discovery which still has a major effect on our lives today. He discovered that a conductor, having no electrical connection with a source of e.m.f. which is moved inside a magnetic field will have a potential difference between its ends.

This was how the principle of electrical generation was born. All modern development has stemmed from this basic discovery. The generation of all the electricity used today relies on this very same principle.

The Factors Controlling the Generated e.m.f.

Having made the basic discovery, Faraday set about trying to discover how to make the generation process more effective. He first attempted to find out what factors affected the size of the e.m.f. generated. The whole process became known as electromagnetic induction, and the e.m.f. generated was therefore known as the induced e.m.f.

All students of electrical principles have to learn Faraday's laws of electromagnetic induction, but first let us examine, step by step, the factors which Michael Faraday considered. The laws will thus be easier to understand and appear much more logical.

Faraday first discovered that a piece of wood or glass placed in the magnetic field had no e.m.f. induced along it. His conclusion was that there must be some form of conductor involved to produce an induced e.m.f.

Secondly, it was found that the e.m.f. disappeared altogether when the conductor was stationary in the field. This suggested that a relative movement must take place between the conductor and the field. If either the conductor or the field moved an e.m.f. would result, but when no relative movement took place, there was no induced e.m.f.

It then seemed a natural step to change the direction of movement and see what effect this produced. The conductor was moved in three different directions and the results compared. Figure 37 demonstrates the possibilities.

It was discovered that the maximum value of induced voltage resulted from the movement shown in (*a*), whereas (*b*) gave a reduced value, and (*c*) none. The conclusions were important. The value of the induced e.m.f. was always a maximum when the conductor cut the field at right angles. If the angle between the direction of movement and the magnetic lines of force is progressively reduced, then the induced e.m.f. is also reduced. Eventually, when the conductor is moved along in the direction of the field, the induced e.m.f. falls to zero.

Faraday could now add the second piece of the jigsaw. The conductor had to cut the lines of force when it moved. A movement in the direction shown in Fig. 37(*d*) induced an e.m.f. equivalent to the motion shown in Fig. 37(*a*) but of the opposite polarity. This shows that the direction of the induced e.m.f. can be reversed by reversing the direction of movement of the conductor. This effect was later investigated by the German scientist Lenz, who gave his name to the law explaining it.

Next it seemed sensible to examine whether moving the conductor at different speeds had any effect on the e.m.f. It was discovered that the faster the conductor moved, the larger was the induced e.m.f. This suggested that it was the rate at which the lines of force were being cut which was the important factor.

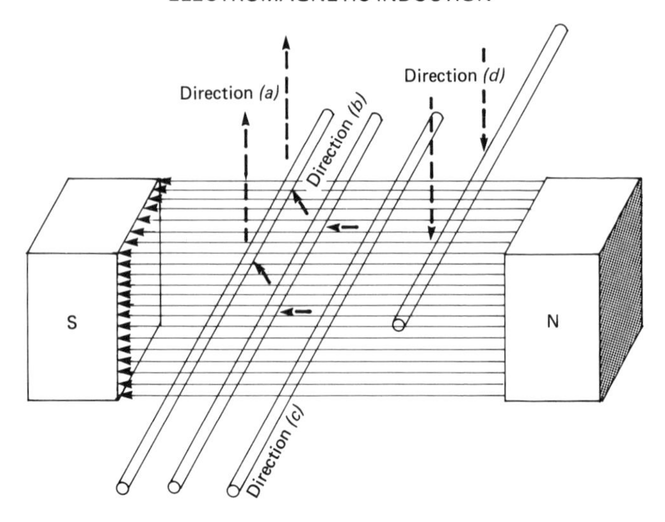

Fig. 37. *Conductor movement in a magnetic field*

Indeed, this indicated two methods of increasing the induced e.m.f. The first would be to move the conductor quickly; the second would be to increase the number of flux lines by increasing the flux density $\boldsymbol{B}$, then a movement of the same speed, would result in a larger induced e.m.f.

It was also important to examine whether or not all of the conductor must be inside the field to be effective. Figure 38 shows the possibilities.

A conductor of the length shown in (*b*) will have a noticeably larger e.m.f. induced in it than one of the length shown in (*a*). A longer conductor, such as that shown in (*c*), has no greater e.m.f. induced in it than one of the length shown in (*b*). The following conclusions can therefore be drawn.

The (*a*) conductor will cut fewer flux lines than (*b*) because of its size, and produce a correspondingly smaller e.m.f. However, the shaded sections of (*c*) are outside the magnetic field and hence will not cut lines of flux. They are thus non-effective, and the best length is obviously a conductor which just stretches across the full magnetic field.

Faraday's Laws

All of these points can now be drawn together into two sentences, which very cleverly cover all of the factors which we have considered. These are known as "Faraday's laws of electromagnetic induction". They should be understood and remembered.

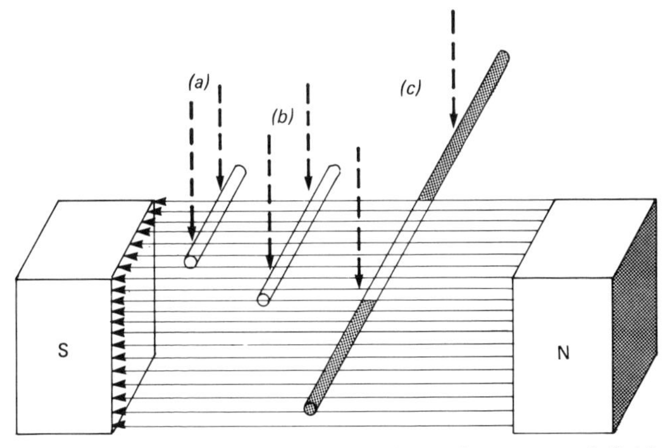

Fig. 38 *Conductors of different lengths cutting a magnetic field*

I. Whenever a conductor cuts lines of force, an e.m.f. will be induced in it.

II. The size of this induced e.m.f. depends upon the rate at which the lines of force are being cut.

The Simple Expression for the Size of the Induced e.m.f.

The factors which we have considered can all be shown in the simple expression: $e = \mathbf{B}lv$. Where: e is the induced e.m.f. in volts; $\mathbf{B}$ is the flux density in tesla; l is the length of conductor in the magnetic field in metres; and v is the velocity of the moving conductor perpendicular to the direction of the lines of force in metres per second.

The various symbols represent those factors explained in the previous paragraphs, and all such expressions, are easier to remember if the reasons behind them are understood.

Example

Calculate the p.d. induced between the ends of a conductor 10 cm long, which cuts at right angles across a magnetic field of flux density 0.4 T, if all the conductor is within the magnetic field and it moves with a velocity of 3.5 m/s.

From $$e = \mathbf{B}lv$$

$$e = 0.4 \times \frac{10}{100} \times 3.5$$

$$e = 0.14\ \text{V}.$$

Looking again at this simple formula, it is easy to appreciate that if the flux or the speed doubled or halved, then the induced voltage would also double or halve. The relationship between these quantities is said to be linear. The whole expression is also described as being linear.

Fleming's Right Hand Rule

It is very useful at this point to have a simple rule when trying to remember in which direction the e.m.f. is induced. Such an aid is called a mnemonic and Fleming's right hand rule is a mnemonic which is very widely known in electrical engineering.

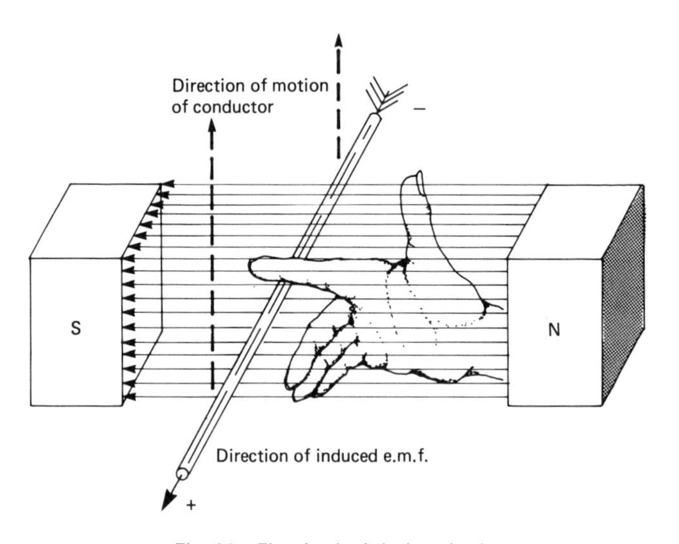

Fig. 39. *Fleming's right hand rule*

Since, by convention, the magnetic field is assumed to have a direction from north pole to south pole of the magnet which produces it, and the conductor moves in a particular direction, the direction of the induced e.m.f. must relate to them in some way. Fleming used the thumb, first finger and second finger of his right hand and held them mutually at right angles. This can be seen in Fig. 39, and it can be appreciated that when any two of these directions are known, the third can readily be deduced by aligning the thumb or fingers with the known quantities and observing the direction of the third digit.

THE GENERATOR PRINCIPLE

Once Faraday's laws had been discovered, engineers developed the principles to the point where they could be practically applied. It was impossible to collect separate conductors after they had moved through a magnetic field and feed them back in again at the top. The only practical method was to shape the conductors into the form of a continuous loop and then rotate the loop within the magnetic field. Figure 40 shows the basic idea. In

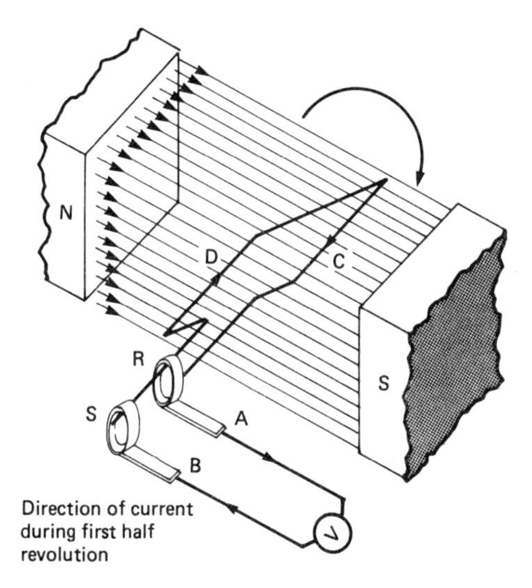

Fig. 40. *E.m.f. induced in a rotating coil*

this way the coil can be continuously moving within the magnetic field. This is the basis for all the rotating generators which we have today. However, the generated voltage is of a particular form. The e.m.f. can be seen to vary between a maximum value and zero as the coil was kept rotating. The polarity of the e.m.f. then reverses, and again a maximum value achieved before returning to zero. This type of varying e.m.f. is known as *alternating* and is always produced by a rotating generator.

An Explanation of the Generated Waveform

The five separate sections of Fig. 41 represent five intermediate positions of the rotating coil as it turns through 360°. It can be assumed that it is rotating at a constant speed in a clockwise direction.

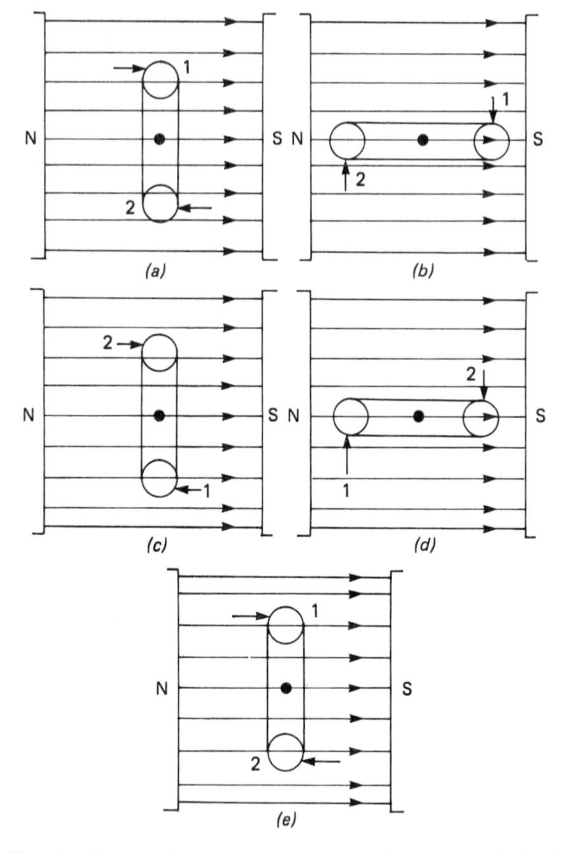

Fig. 41. *Coil rotating at constant speed in a magnetic field*

The conductor movement in position (*a*) is along the direction of the field, and when the coil is absolutely vertical, there is no induced e.m.f. Between positions (*a*) and (*b*), for a given rotational distance moved, more and more lines are being cut and the e.m.f. increases. At position (*b*), the movement of the coil is at right angles to the field, and the e.m.f. induced is a maximum. Between positions (*b*) and (*c*) the e.m.f. falls, eventually reaching zero when it is at position (*c*).

Between positions (*c*) and (*d*), side 1 of the coil, which has been moving down through the field, now starts to move upwards and the polarity of the induced e.m.f. is reversed. Once again the value of the e.m.f. will vary, reaching a maximum in the position (*d*), before falling back to zero as the coil completes its full revolution and returns to position (*a*).

These changes in the induced e.m.f. are plotted in the Fig. 42 and produce what is known as a sine wave, because it has the same shape as a graph of the sine function.

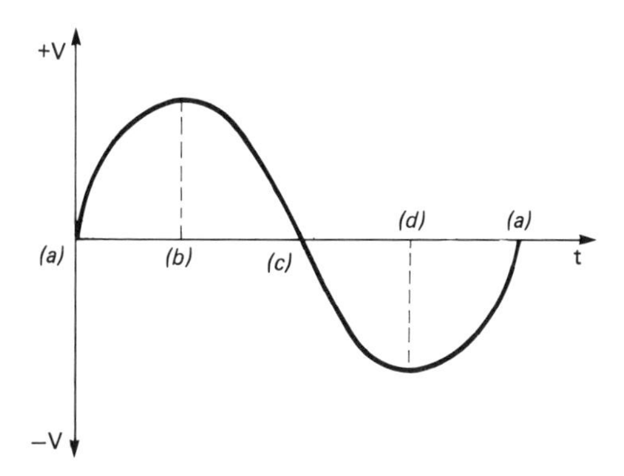

Fig. 42. *Curve produced by plotting the variation in e.m.f. induced in the coil in Fig. 41*

Lenz's Law

Lenz expanded the early work on induction in an effort to establish the laws which governed the polarity of the induced e.m.f.s. He concluded that whatever effect brought about the induction, the generated e.m.f. would tend to oppose it. The law, known as Lenz's law, states just that:

The direction of the induced e.m.f. is such that it opposes the effect producing it

This is common sense if a little thought is given to it. Natural laws do not give us "something for nothing", and if a conductor is being moved we have to do work to produce the movement. We produce electrical energy by expending mechanical energy. Part of the energy is used up in overcoming forces set up by the induced e.m.f.

THE MOTOR PRINCIPLE

One of the first rules which we learn in magnetism is that like poles repel. Take two simple bar magnets and place the two north poles close to each other. You will feel a physical repulsion

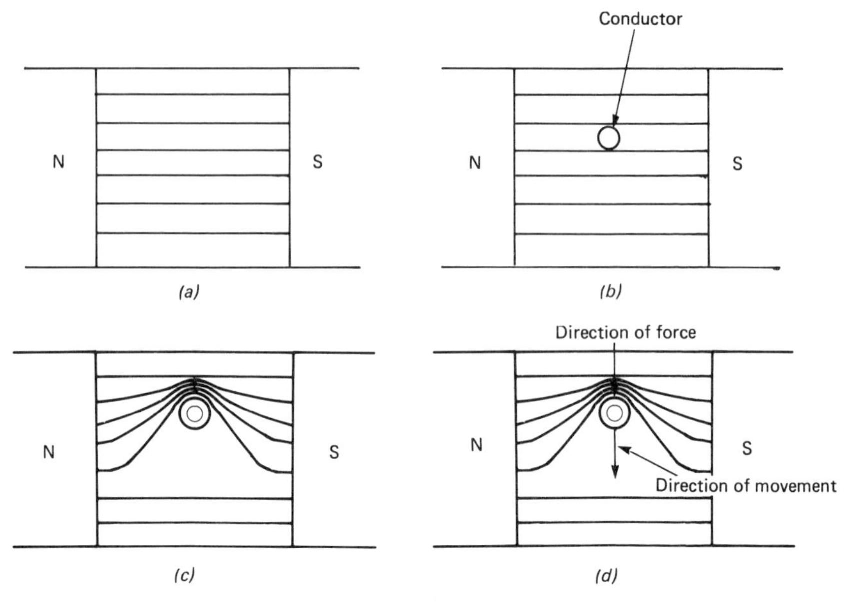

Fig. 43. *Force exerted on a current-carrying conductor in a magnetic field*

between them. This is caused by the reaction of the two magnetic fields around the two permanent magnets.

As we know, a magnetic field exists between the poles of a permanent magnet such as that shown in Fig. 43(*a*).

When a conductor is introduced into the field, no reaction takes place and the field pattern remains undisturbed. Now let us pass a current through the conductor. We know from past experience (*see* Fig. 11) that this will produce a circular magnetic field around the conductor.

In just the same way as the fields of two permanent magnets reacted, so will the fields produced by the permanent magnet and by the single conductor. This produces the field pattern shown in (*c*). It can be seen that the new field pattern is not uniform. A heavy concentration exists at a point above the conductor, and a null point below it. All physical laws attempt to maintain or restore an equal balance, and in this case the conductor is subjected to a force which attempts to reject it from the field and thus allow the field to once again become uniform. In this example the force is downwards as shown in (*d*), and the conductor will move if it is free to do so.

This establishes a second very important principle. The application of electrical energy to a conductor which is within a magnetic field will produce a mechanical force. Indeed this is the reverse of the principle of the generator, and it is the basic principle on which all electric motors are based. It is called *the motor principle.*

The Factors Which Influence the Size of the Force

Since the motor force is produced by the reaction of two magnetic fields, it seems perfectly reasonable that if we increase the strength of one or both of them, the resultant force will also increase. This is quite true and an increase in the flux density of either field will produce a larger force. By passing a larger current through the conductor the magnetising force will increase, and so will the flux density associated with it. Thus the force is proportional to the current.

The size of the conductor will also influence the force produced. A conductor which has only half of its length in a magnetic field will not generate as great a force as one of the same length which is fully immersed in it.

The Expression Governing the Size of the Motor Force

All three factors can readily be expressed in a simple linear expression, i.e. the force will vary proportionately as each component of the expression varies. The force is expressed simply as: $F = \boldsymbol{B}lI$ where F is the force produced in newtons; $\boldsymbol{B}$ is the flux density of the ambient magnetic field in teslas; l is the active length of the conductor in metres; and I is the current flowing through the conductor in amperes.

Example 1

Calculate the force in newtons exerted on a conductor of length 14 cm if it is all immersed in a magnetic field of flux density 1.2 T and a current of 0.5 A flows through it.

$$\begin{aligned} \text{From } F &= \boldsymbol{B}lI \\ &= 1.2 \times 0.14 \times 0.5 \\ &= 0.084\,\text{N}. \end{aligned}$$

An Explanation of the Lenz Principles

It is important at this stage to appreciate the effect of Lenz's law. The conductor, when subjected to the motor force, will move in the field. We have already seen, from the laws of Faraday that wherever a conductor cuts lines of force, an e.m.f. will be generated in it. Application of Fleming's right hand rule, shows that the direction of this induced e.m.f., is such that it opposes the applied voltage. This tends to reduce the current flowing which in turn from $\boldsymbol{B}Il$ tends to reduce the force produced.

Lenz's law is therefore justified. The direction of any induced e.m.f. is such that it tends to oppose the effect producing it. In this case it is the movement.

The Direction of Force

Further reference to the diagram Fig. 39 shows that for the given conditions the movement of the conductor will be in a particular direction. It is very important that we are able to decide which direction this will be. Once again a simple mnemonic comes to our aid.

A second Fleming's rule, very similar to the right hand rule for generators, is used. In this case the thumb, first and second fingers of the left hand are held mutually at right angles, and represent the direction of the motion, field and current respectively. By pointing the first and second fingers in the direction of the known quantities namely the field and current, the unknown direction of movement can be found by noting the direction pointed out by the thumb. Figure 44 shows the application of this left hand rule.

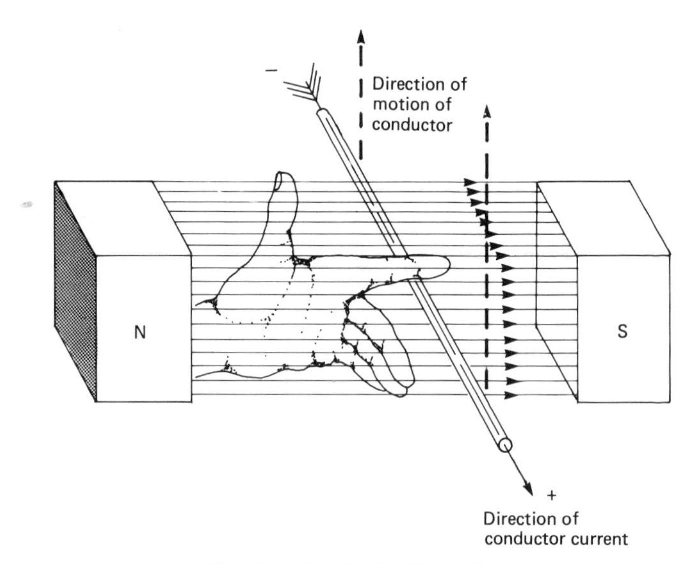

Fig. 44. *Fleming's left hand rule*

SELF-INDUCTANCE

So far we have carefully considered the principles of generation and of electric motors. In each case we have looked at the principle, effects and how these can be best applied. Both these are on the useful side and our life today would be very different if we did not have the generated electric power, or the electric motors to make use of that power. Unfortunately, not all of the effects investigated by Michael Faraday were useful and one which often causes problems in electrical engineering is the effect known as self-inductance.

Let us think for a moment of a simple coil of wire of a few turns, connected to a d.c. supply. We know that when the current is switched on, the magnetic field, which you remember consists of concentric circles around each conductor, will begin to increase. As the field increases, the concentric lines of force expand and eventually cross adjacent turns of the coil. The word "whenever" in Faraday's laws becomes very important. Whether we want it or not, an e.m.f. will be generated in the adjacent conductors. You may think that this is fine, that it will help to drive the main current through the circuit but from Lenz's law we can see that it is generated in such a direction that it tends to oppose the effect producing it.

The effect producing the e.m.f. in this case is the initial setting up of the current. Therefore, the direction of the induced e.m.f. is in direct opposition to the applied e.m.f. and the current is held down below the value which we would expect from applying Ohm's law.

You might well wonder if Ohm's law ever applies when a coil is connected in a d.c. circuit. Ohm's law does apply, but only after a short time (of the order of milliseconds) has elapsed. The explanation is again found in the application of Faraday's laws. The conductor must *cut* the magnetic field. In the example which we are investigating, the field is the moving part, cutting the adjacent turns of the coil. However, the field will reach a maximum value around the conductor and will then become stationary. When this occurs the lines of force will no longer be cut by the adjacent turns of the coil and thus there will be no back e.m.f. induced. Therefore, when the field has reached its full value, there is no relative movement and hence no e.m.f. in opposition to the e.m.f. from the d.c. supply. The current will reach the value expected from Ohm's law.

The buildup of the magnetic field is not linear but tends to be much faster at first. This means that the rate at which lines of force are cut will be higher, and the induced e.m.f. will be a maximum. As the rate of buildup slows down, the induced e.m.f. will reduce giving less opposition to the applied voltage. Eventually, the current reaches a value wholly dependent on the amount of resistance in the circuit and the e.m.f. of the d.c. supply, and the induced e.m.f. reduces to zero. The circuit now conforms to the basic principle of Ohm's law.

This is clearly shown by drawing a graph to represent the buildup of current from when the source of e.m.f. is first connected to the time when the value expected from Ohm's law is reached (*see* Fig. 45).

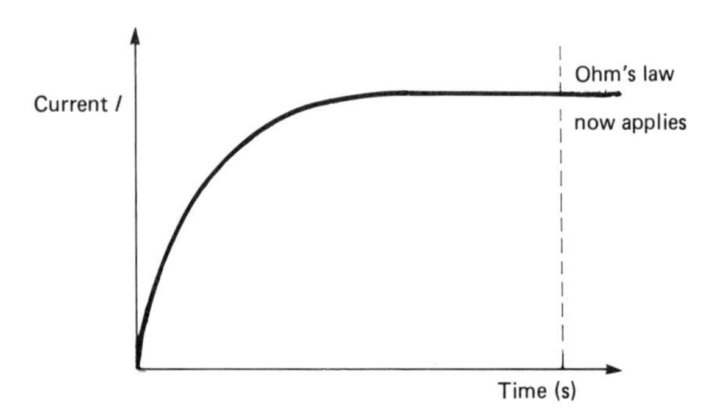

Fig. 45. *Growth of current in an inductive circuit*

It can be appreciated that this delay in reaching the ohms law value can be very important to the engineer. The effect, which is known as self-induction for obvious reasons, has also led to coils being known as inductances.

Engineers are resourceful people, however, and it must be mentioned that what is a problem in some circuits is deliberately made use of in others. The radio engineer, for example, has many applications for inductance.

The Unit of Inductance

As usual, to make sense of this new effect we must be able to measure it, so that one inductance can be compared with another. To measure anything we need units, and in this case the unit is the *henry* (symbol H). Inductance is to do with the voltages which are generated when changes occur in a magnetic field. The magnetic field changes because the current is made to change by changes in the circuit conditions. These facts are tied together in the definition of the henry.

When a current in a conductor changes at one ampere per second and in doing so causes a back e.m.f. of one volt, then that conductor is said to have an inductance of one henry.

The Simple Expression Governing Inductance

In our simple explanation of inductance, we used the idea that the current associated with one turn of conductor in a coil builds a magnetic field, which inevitable cuts the turn next to it. In a practical situation it will also cut the other turns in the coil. Each of these in turn will also be producing a magnetic field which will cut each of the other turns and so on. The idea, then, is that we can produce an expression which combines the flux and the number of turns. Another word for turn of wire is a link and the amount of inductance a coil has will depend on the number of turns or links and the flux. Not only the amount of flux is important but, as Faraday's second law states, how the flux is changing in relation to the current will also be important.

Inductance can be expressed as being proportional to the change of flux linkage, compared to the change in current.

$$L = N \times \frac{\text{change in flux linkage}}{\text{change in current}}.$$

Now the mathematician sees writing "change in current" and "change in flux" as being extremely clumsy. He writes these expressions simply using $\mathrm{d}\Phi$ and $\mathrm{d}I$ for very small changes in flux and current.

The expression for inductance can therefore be written:

$$L = N \times \frac{\mathrm{d}\Phi}{\mathrm{d}I}\,\mathrm{H}.$$

A simple example will show how this expression is used.

Example 2

A flux of 50 μWb is produced when a current of 5 A is made to flow in a coil of 2,000 turns. What is the inductance of the coil? The change in flux is 50 μWb, i.e. the flux has risen from zero to 50 μWb. The change in current which brought about the change in flux is 5 A, i.e. the current has risen from zero to 5 A. Therefore,

using $$L = N \times \frac{\mathrm{d}\Phi}{\mathrm{d}I},$$

$$L = 2{,}000 \times \frac{50 \times 10^{-6}}{5} = 0.02\ \mathrm{H}.$$

The Expression for Induced e.m.f.

It seems logical that a simple expression will help us calculate the value of the voltage which will be induced in a coil of known inductance.

$$E = L\frac{\mathrm{d}I}{\mathrm{d}t}\,\mathrm{V}$$

L = inductance in henrys
$\mathrm{d}I$ = the change in current
$\mathrm{d}t$ = the change in time.

The expression $\mathrm{d}I/\mathrm{d}t$ is therefore the rate of change of current and exactly fits Faraday's second law. Another example will help to show how this expression is used.

Example 3
When the current flowing through a coil of inductance 5 H rises from zero to 4 A in 0.02 s, what will be the induced e.m.f.?

$$E = L\frac{\mathrm{d}I}{\mathrm{d}t}$$

$$= \frac{5 \times 4}{0.02} = 1{,}000\ \mathrm{V}.$$

ARCING IN INDUCTIVE CIRCUITS

Example 3 shows two things. First how straightforward it is to apply the expression used, and second how large a voltage can be generated by a current of only 4 A.

The power engineer often has to cope with such large induced voltages, because, of course, exactly the same voltage will be generated when the current is switched off. Many circuits which are fed by the mains (such as the windings for motors and the bars of electric fires) are in the form of coils and present considerable inductances. When these circuits are switched off, the large voltages which are induced often cause arcs. You may have seen this yourself as you switch off an electric fire, particularly in the dark. In large installations or power stations arcs have to be avoided where possible and circuit breakers are used to make the switching of highly inductive circuits quite safe.

Calculation of Induced Voltage when the Flux and not the Current is known

A substitution of known factors can often help solve a problem when it could be difficult because some factors are not known.

A good example of such a manipulation is given here. Let us assume that the value of the induced voltage is needed, but the flux change rather than the change in current is known. We know that;

$$E = L\frac{dI}{dt},$$

and that $$L = N\frac{d\Phi}{dI}\,\text{H}.$$

By substituting for L we get;

$$E = N\frac{d\Phi}{dI} \times \frac{dI}{dt}$$

$$\therefore \quad E = N\frac{d\Phi}{dt}\,\text{V}.$$

This simply means that the value of the induced voltage is given by the rate of change of flux linkages.

Example 4

What is the value of the e.m.f. induced when a flux of 5 mWb associated with a 5,000 turn coil collapses in 0.01 s?

From $$E = N\frac{d\Phi}{dt}\,\text{V}$$

$$E = \frac{5{,}000 \times 5 \times 10^{-3}}{0.01}$$

$$= 2{,}500\,\text{V}.$$

Energy Stored in an Inductor

When the current is switched off in an inductive circuit the value does not fall to zero instantaneously. The induced voltage maintains the current for a short period after the opening of the switch contacts. It is this maintained current which produces the arc across the opening contacts and is a further example of Lenz's law.

In this case the current is collapsing and trying to reach zero, but the inductance effect maintains it. We cannot get something for nothing and we are now disconnected from the source of energy, the electrical supply. The energy to produce the arc must come from somewhere, and indeed it comes from the energy which is stored in the magnetic field around the coil.

The actual amount of energy is dependent upon the value of current which is flowing and the size of the coil, i.e. the number of turns. We shall show that the energy stored = $\frac{1}{2} LI^2$ J.

This expression is derived quite simply and the derivation is a good example of how, using the basic building bricks which were mentioned in Chapter 1, other expressions can be deduced.

Let us take it one step at a time. The amount of energy stored in the magnetic field must be equal to the energy fed to the coil, during the time the field is building. The average current flowing whilst the value rises from zero to I must be half the maximum value, i.e. $I_{Av} = I/2$, where I is the maximum value. The average rate of change of current will depend on how long it takes to change. Average rate of change = I/t, where t is the time taken for the current to rise from zero to its maximum value.

Hence, from e.m.f. the expression shown recently:

$$\text{Average induced e.m.f.} = L \times \text{rate of change of current}$$
$$= L \times \frac{I}{t}.$$

$$\text{Energy} = \text{watts} \times \text{seconds}.$$
$$= \text{volts} \times \text{amps} \times \text{time}$$
$$= L \times \frac{I}{t} \times \frac{I}{2} \times t$$
$$= \tfrac{1}{2} L I^2 \text{ J}.$$

Example 5

Calculate the energy stored in a coil which has a self inductance of 5 H when a current of 2 A is passed through it.

$$\text{Energy} = \tfrac{1}{2} LI^2 \text{ J}$$
$$= \tfrac{1}{2} \times 5 \times 10^{-3} \times 2^2$$
$$= 0.01 \text{ J}.$$

THE INDUCTANCE OF A COIL IN RELATION TO ITS PHYSICAL PROPERTIES

To follow the development of this section it is necessary to summarise some fundamental magnetic circuit principles.

The magnetic force $\boldsymbol{H}$ for a coil is given by NI/l.

Where: N = number of turns;
l = length of magnetic circuit; and
I = current in amperes.

The total flux Φ is equal to the product of the flux density $\boldsymbol{B}$ and the area A over which the flux is set up, i.e.

$$\Phi = \boldsymbol{B} \times A$$

The permeability μ of a medium in which a magnetic field is being established is given by the ratio $\boldsymbol{B}/\boldsymbol{H}$, i.e.

$$\mu = \mu_0 \mu_r = \frac{\boldsymbol{B}}{\boldsymbol{H}}$$

Therefore the flux Φ can be written:

$$\Phi = \boldsymbol{B} \times A.$$
$$= \mu_0 \mu_r \boldsymbol{H} \times A.$$

Substituting for $\boldsymbol{H}$,

$$\Phi = \frac{\mu_0 \mu_r NIA}{l}.$$

Now we can develop an expression for the inductance of the coil in terms of these physical properties, and we know that:

$$L = \frac{N\Phi}{I}.$$

Therefore, $$L = \frac{N\mu NIA}{I \times l}$$
$$= \frac{N^2 A \mu_0 \mu_r}{l} H,$$

which is the same as N_2/S.

We can deduce several things from this expression, all of which are important when an engineer is attempting to design a coil. Remember that some designs will need the minimum inductance, whereas others require the coil to have a large inductance.

We can now see that:

(*a*) an increase in the number of turns will give a large increase in the inductance, since it is proportional to the square of the number of turns;

(*b*) the cross-sectional area of the coil must be as small as possible if the inductance is to be kept low (using the expression above, when A doubles so does the inductance);

(*c*) when the number of turns is concentrated over a small length the inductance will increase (e.g. if l is halved then the inductance is doubled); and

(*d*) the inductance can be increased by putting a core in the coil which has a high permeability, because when the coil is air cored $\mu = \mu_0 \times 1$.

However, by putting an iron core inside the coil the inductance value can be made several hundred or even several thousand times larger. To the radio engineer this may be very important. He often needs high values of inductance but at the same time requires small physical components.

MUTUAL INDUCTANCE

Let us recapitulate the basic principle of inductance. It is the *moving* magnetic field, which cuts other turns of the coil and induces an e.m.f. into those turns. It seems quite reasonable that if this moving field cuts the turns of a second coil, an e.m.f. will also be induced into that second coil.

This is indeed the case. An e.m.f. will appear across the ends of the coil B in Fig. 46 but only when the magnetic field of coil A is being changed.

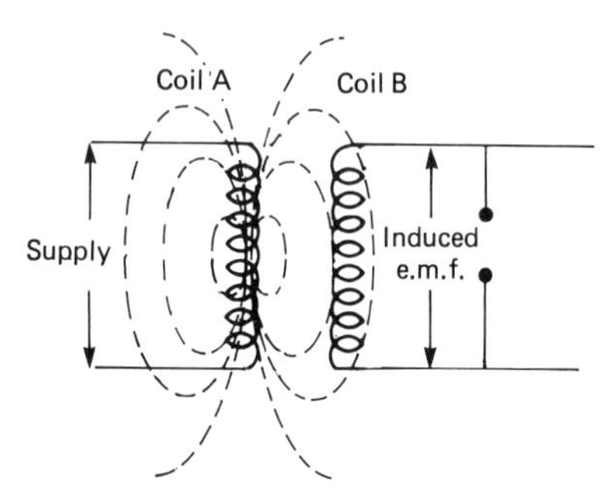

Fig. 46. *Induced e.m.f. caused by mutual inductance effect*

The ends of coil B can now be joined and a current will flow. This then is a very useful engineering tool. The coils are not in any way physically connected together, but a current can be made to flow in coil B by driving a current through coil A.

As usual, there is a snag. The current in coil B will only flow during the first few milliseconds. Once the field in coil A is established it is no longer moving and the induced e.m.f. will be zero.

The effect of a current changing in one coil producing an induced e.m.f. in a second coil is known as mutual inductance.

The Unit of Mutual Inductance

This is also the henry but the basic definition will be slightly modified to:

> When a current changing at the rate of 1 ampere per second in one coil, causes an e.m.f. of 1 volt to be induced in a second coil, then the two coils are said to have a mutual inductance of 1 henry.

THE BASIC PRINCIPLE OF THE TRANSFORMER

The usefulness of a current which endures for only the first few milliseconds is likely to be very small, but if the current in coil A is made to vary continuously, the field will also be changing and an e.m.f. will be induced in coil B all the time.

This is the basic principle of the transformer. Two coils are wound and mounted close to each other. An alternating voltage is fed to the first coil, which we will call coil A. The current driven through coil A will change continuously, hence the magnetic flux associated with it will be permanently on the move. This will produce an e.m.f. in coil B which will follow the same changing pattern. The transformer is a most useful device for all kinds of electrical and electronic engineers.

The Voltage or Turns Ratio

Although it would be very difficult to achieve in practice, for the sake of explanation let us assume that we could produce one single line of magnetic flux from the first coil. As this line moves across a turn of the second coil, then it will induce a certain voltage in it. It will also induce this voltage across a second, a third, a fourth, a fifth turn, and so on. The number of turns in the second coil is therefore very important. The greater the number, the larger the total e.m.f. available at its terminals. We already know that the amount of flux which is set up in coil A will be proportional to the number of turns in it. The ratio of the number of turns in coil A to coil B is therefore the basic controlling factor of the transformer. The respective coils are often called primary and secondary coils. Since the primary magnetising current will be proportional to the applied voltage, the ratio of the primary e.m.f. to the output or secondary e.m.f.

can be expressed in terms of the number of turns on the respective coils.

This is often called the voltage or turns ratio and is expressed:

$$\frac{N_p}{N_s} = \frac{E_p}{E_s}.$$

Where: N_p = number of turns in the primary coil;

N_s = number of turns in the secondary coil;

E_p = primary e.m.f.; and

E_s = secondary e.m.f.

Example 6

The primary winding of a simple transformer has 1,000 turns, what will be the output voltage if the input voltage is 25 V and the secondary has 4,000 turns.

$$\frac{E_p}{E_s} = \frac{N_p}{N_s}$$

Therefore,

$$\frac{E_s}{E_p} = \frac{N_s}{N_p}$$

Therefore,

$$E_s = \frac{N_s}{N_p} \times E_p$$

$$= \frac{4{,}000}{1{,}000} \times 25\text{ V}$$

$$= 100\text{ V}.$$

The transformer can be used for increasing or decreasing the voltage and is widely used for these purposes.

SELF-ASSESSMENT QUESTIONS

1. A conductor 1.2 m long cuts a magnetic field of flux density 1.5 T at right angles. If the whole length of the conductor cuts the field at a velocity of 6 m/s, calculate the induced e.m.f. in the conductor.

2. A coil has a total active conductor length of 8.5 m and is completely within a field of flux density 1 T. If the coil is rotated so that the active length cuts the field at right angles at 200 rev/min, calculate the e.m.f. induced if the coil diameter is 30 cm and the radius about the axis of rotation is 15 cm. If the coil ends

are connected to a load of resistance 10 Ω calculate the current flowing in the coil and hence the value of the force required to rotate the coil.

3. If a current of 4 A produces a flux of 4.8 μWb in an air cored coil of 1,600 turns calculate the inductance of the coil.

4. A current of 2.5 A flows through an air cored coil of 1,000 turns. If the coil inductance is 0.6 H, calculate the magnetic flux.

5. A coil of 2,000 turns is uniformly wound on a wooden ring having a mean diameter of 32 cm and cross-sectional area of 4 cm^2. Calculate the inductance of the coil.

6. An electromagnet having an inductance of 2.5 H has a current flowing through its winding of 1.2 A. Calculate the stored energy.

7. A current flowing through a coil of inductance 16 mH is reversed over a period of 20 ms and the e.m.f. induced is 8 V. Calculate value of current.

8. A coil of 2,000 turns has a flux of 10 mWb when the current flowing in it is reversed in 0.05 sec. Calculate the e.m.f. induced in the coil.

9. Calculate the inductance of an iron cored coil of effective length 31.4 cm and cross-sectional area 5 cm^2 which is wound with 1,500 turns, if the relative permeability of the iron under these conditions is 1,000.

10. Two coils A and B, having 1,000 and 800 turns respectively, share a common core and all the flux produced by one coil links with the other.

If the value of flux in the core is 3 mWb and this collapses in 0.04 s, calculate the e.m.f. induced in each coil.

11. A transformer which has 3,000 turns on the primary winding has a supply voltage of 400 V applied to it.

If the induced voltage in the secondary winding is 100 V, calculate the number of secondary turns.

CUMULATIVE QUESTIONS, CHAPTERS 1–4

1. The current through the meter in the diagram must be:

(*a*) 4 A;
(*b*) 1 A;
(*c*) $\frac{2}{3}$ A;
(*d*) 6 A.

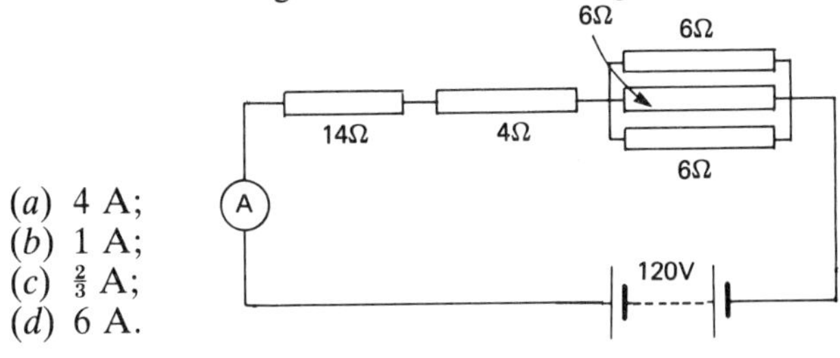

2. The voltage across the 6 Ω resistor must be:

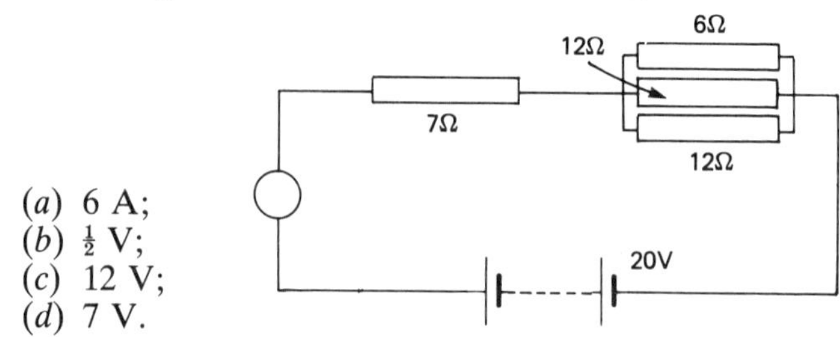

(*a*) 6 A;
(*b*) $\frac{1}{2}$ V;
(*c*) 12 V;
(*d*) 7 V.

3. Calculate:
   (*a*) the currents through the 20 Ω and 30 Ω resistors;
   (*b*) the voltage across the whole circuit; and
   (*c*) the total power.

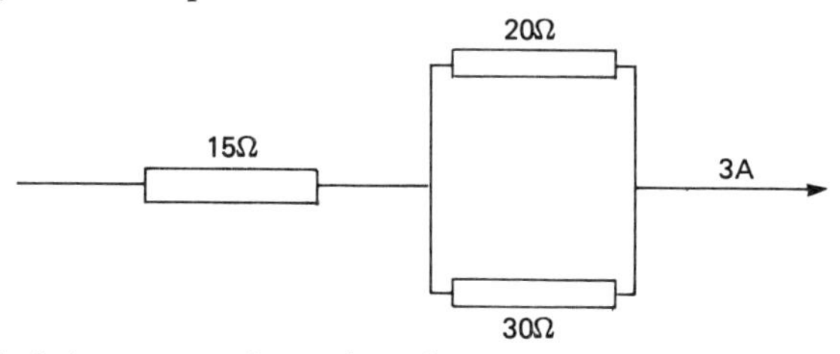

4. Find the current in each resistor.

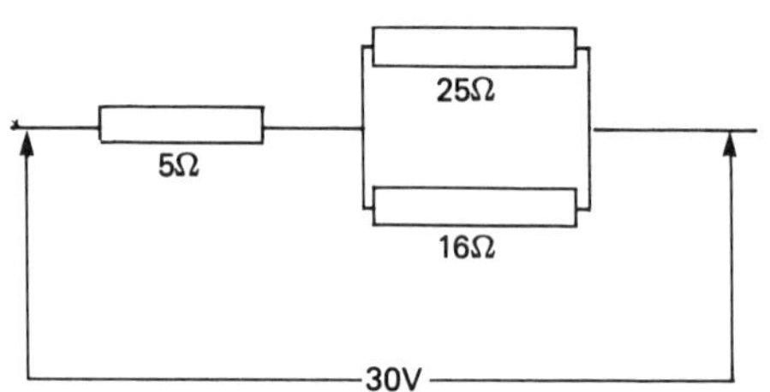

5. Two resistances A and B, of 20 Ω and 28 Ω respectively, are connected in parallel across a battery of 20 cells. Each cell has an e.m.f. of 2.1 V and an internal resistance of 0.13 Ω. Calculate:
   (*a*) the current produced by the battery; and
   (*b*) the currents through A and B.

6. Solve the circuit.

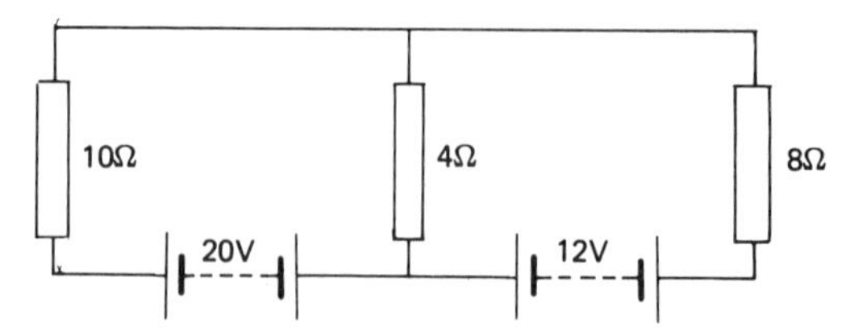

7. Solve the circuit.

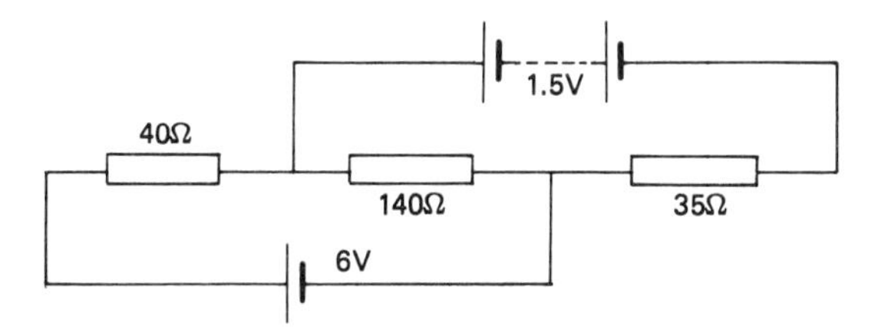

8. Using the superposition theorem and then Kirchhoff solve the following circuit.

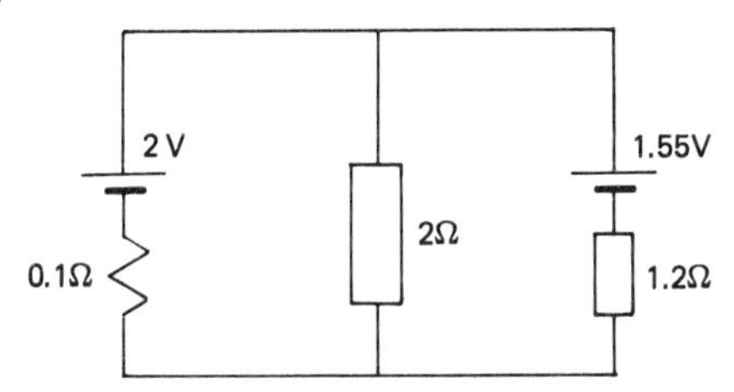

9. An air-cored coil of inductance 4.5 mH has an axial length of 314 mm and a cross-sectional area of 500 mm^2. How many turns are there in the coil?

10. A current of 2.5 A flows through a 1,000-turn coil that is aircored. The coil inductance is 0.6 H. Calculate the magnetic flux produced.

11. A toroid of mean diameter 200 mm has a uniformly wound conductor passed ten times through its centre. The cross-sectional area of the ring is 100 mm^2 and its relative permeability is 850. A current of 1.5 A is passed through the conductor. Calculate the flux in the ring.

12. A coil of insulated wire of 400 turns and of resistance 0.25 Ω is wound tightly round a steel ring and is connected to a d.c. supply of 4.0 V. The steel ring is of uniform cross-sectional area 600 mm^2 and of mean diameter 150 mm. The relative permeability of the steel is 450. Calculate the total flux in the ring.

13. A steel ring has a circular cross-sectional area of 150 mm^2. The ring has a mean diameter of 85 mm and is wound with 250 turns of wire. The steel has a relative permeability of 500 and the total flux in the steel is 0.35 Wb. What current is required in the coil to produce this flux?

14. A 2,000-turn coil is uniformly wound on an ebonite ring of mean diameter 320 mm and cross-sectional area 400 mm^2. Calculate the inductance of the toroid so formed.

15. Three capacitors of capacitance 2 μF, 3 μF and 6 μF respectively are connected in series across a 500 V d.c. supply. Calculate:

(*a*) the charge on each capacitor
(*b*) the p.d. across each capacitor
(*c*) the energy stored in the 6 μF capacitor.

16. Two capacitors, having capacitances of 10 μF and 15 μF, are connected in parallel and a capacitor of 8 μF is connected in series with them. What single capacitor would replace all three to give the same value of capacitance?

17. The energy stored in a certain capacitor when connected across 400 V d.c. is 0.3 J. Calculate:

(*a*) the capacitance;
(*b*) the charge on the capacitor.

18. Two capacitors with capacitances of 10 μF and 2 μF are connected (*a*) in series (*b*) in parallel. Find the equivalent capacitances in each case.

19. The total capacitance of two capacitors connected in series is 3 μF. One capacitor has a capacitance of 4 μF. What is the capacitance of the second?

20. Two capacitors, with capacitances 7 μF and 3 μF, are connected in parallel. A third capacitor C is connected in series giving a total capacitance of 6 μF. Find the capacitance of C.

21. A capacitor A is connected in series with two capacitors B and C which are connected in parallel. If the capacitances of A, B, and C are 4 μF, 3 μF and 6 μF respectively, calculate the equivalent capacitance of the combination, If a p.d. of 20 V d.c. is maintained across the whole circuit, calculate the charge on the 3 μF capacitor.

22. A capacitor is made with 7 metal plates separated by sheets of mica 0.3 mm thick and of ε_r 6. The area of one side of each plate is 500 cm^2. Calculate the capacitance in micro-farads.

23. Show from first principles that the total capacitance of two capacitors having capacitances C_1 and C_2 respectively, connected in parallel, is $C_1 + C_2$.

A circuit consists of two capacitors A and B in parallel connected in series with another capacitor C. The capacitances of A, and B and C are 6 μF, 10 μF and 16 μF respectively. When the circuit is connected across a 400 V d.c. supply calculate:

(*a*) the potential difference across each capacitor;
(*b*) the charge on each capacitor.

24. A parallel plate capacitor has a capacitance of 300 pF. It has 9 plates of 4 cm × 3 cm, separated by mica having a dielectric constant of 5. Find the thickness of the mica.

25. If a current of 4 A produces a flux of 4.8 μWb in an air cored coil of 1,600 turns. What is the inductance of the coil?

26. A current of 2.5 A flows through a 1,000 turn, air cored coil. If the coil inductance is 0.6 H find the magnitude of the magnetic flux.

27. When carrying a current of 1.2 A each field coil of a four-pole d.c. generator has an inductance of 2.5 H. What amount of energy is stored in the magnetic field?

28. The current in a coil changes from 20 A to 4 A in 0.4 s and the e.m.f. induced is 50 V. What is the inductance of the coil?

29. A current flowing through a coil of inductance 16 mH is reversed in a time of 20 ms and the e.m.f. induced is 8 V. Calculate the value of the current.

30. A current of 15 A flows through a coil of inductance 60 mH. If the circuit is opened in 15 ms calculate the power to be dissipated.

(Assume that the transformers for questions 31–36 are 100% efficient.)

31. A transformer has a stepdown ratio of 20:1. Its primary winding consists of 2,000 turns. Calculate.

(*a*) the number of turns in the secondary winding; and

(*b*) the secondary voltage when the primary is fed with 240 V.

32. A single phase transformer 240/50 V supplies 150 A from its secondary side. Calculate the primary current.

33. A 240/50 V step down transformer is supplying a current of 80 A. Calculate the primary current.

34. A transformer is required to supply 100 V from 240 V mains. Its primary winding has 2,000 turns. Calculate (*a*) the number of secondary turns; and (*a*) the primary current when the transformer delivers a current of 15 A.

35. The transformation ratio of a transformer is 6:1. Calculate the secondary voltage for a primary voltage of 415 V.

36. Describe the operation of a single phase transformer, explaining the effects of the various parts. The primary winding of a stepdown transformer takes a current of 22 A at 3,300 V when working a full load. If the stepdown ratio is 14:1, calculate the secondary voltage and current.

CHAPTER FIVE

Alternating Voltages and Currents

CHAPTER OBJECTIVES

After studying this chapter you should be able to:

* understand the fundamental properties associated with alternating voltages and currents;
* define the terms and represent alternating quantities by mathematical, graphical and phasor means;
* appreciate the principles of half-wave and full-wave rectification and how simple rectification circuits are constructed.

MATHEMATICAL CONSIDERATIONS

Angular Measure

Let us look at some simple mathematical factors which will help in understanding the detail of this chapter.

First let us consider the measurement of angles. Most people are quite familiar with the measurement of angles in degrees. The degree is 1/360th of a full revolution and angles are therefore measured in multiples of the degree. Each degree is sub-divided into 60 minutes and in this way fractions of the degree are measured, e.g. an angle which is half of 23° is 11°30′. A right angle is an angle of 90°.

An alternative angular measure, which may not be anything like as familiar, or indeed may not be known at all, is the *radian*. Simple mathematics teaches us that there is a relationship between the radius and the circumference of a circle, such that the circumference = $2\pi \times$ radius.

In Fig. 47 we can see that if we mark off part of the circumference of equal length to the radius (shown here dashed) this arc will subtend an angle at the centre. Now if we continue to go around the circle marking off r every time we must eventually mark off 2π arcs of length r, because all the whole circumference is $2\pi\ r$, therefore there must be $2\pi\ r/r = 2\pi$ sections. The angle traced out at the centre in Fig. 47 is said to be 1 *radian*. There are therefore 2π radians around the full 360°, i.e. 2π radians = 360°. The radian is thus simply an alternative measurement of angle,

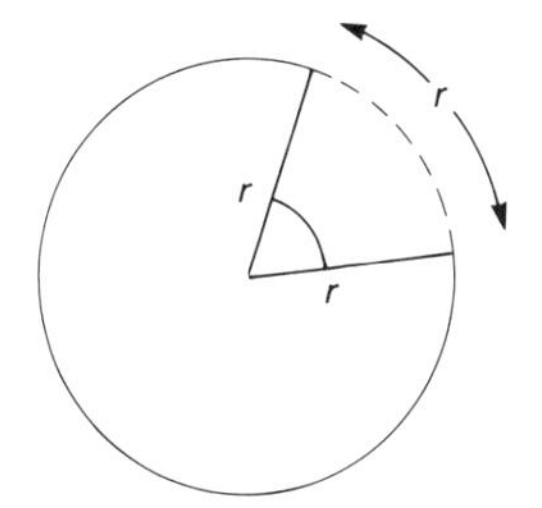

Fig. 47. *The definition of the radian*

and is very often used in electrical engineering. As you will see later on in this chapter, certain angles have a particular significance in alternating current circuits, and it is useful to be able to recognise these angles in both types of measure. It follows from what has been said that $180° = \pi$ radians and $90° = \pi/2$ radians. N.B. $1° = \pi/180$ radians, and 1 radian $= 180°/\pi$.

The Mid-ordinate Rule

Occasionally it is necessary to calculate the area enclosed between a given graph and its axis. More advanced mathematics can help us do this quite easily, but other simple techniques can help us make good approximations, and are suitable for our purposes. The mid-ordinate rule is one such technique.

One of the first calculations which we learn in mathematics is to calculate the area of a rectangle $l \times b =$ area (*see* Fig. 48 (*a*)).

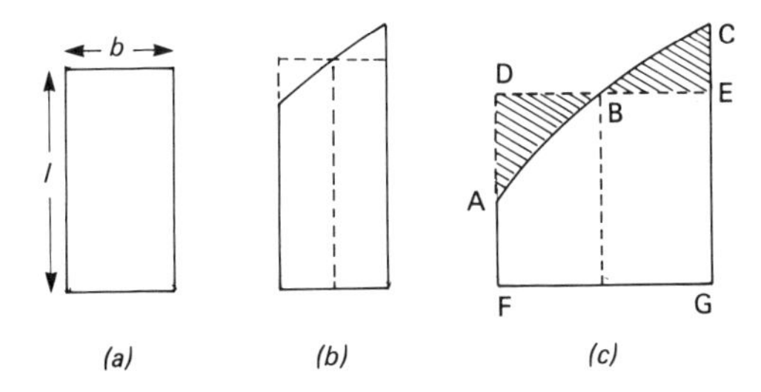

Fig. 48. *Illustration of the use of the mid-value as an approximation*

Now let us assume that we wish to try and calculate the area of the figure in Fig. 48 (*b*). This is a little more complicated because the lengths of the sides are not equal. However, if we examine the top a little more carefully as in Fig. 48 (*c*), we can see that the triangle BEC is an extra area added on to the rectangle FGED.

Triangle ADB is just the opposite, it is missing from the total area of FGDE. Now ADB ≏ BEC and we find that if we calculate the full area using FD × DE then the error which is introduced is quite small and the answer calculated is acceptably accurate. This is exactly the principle used to find the area between a graph and its axis. Assume a graph having the same shape as that shown in Fig. 49.

The graph is split into several separate rectangles of equal width. The height of each rectangle is made to be equal to the length of a line which extends at right-angles to the x-axis from

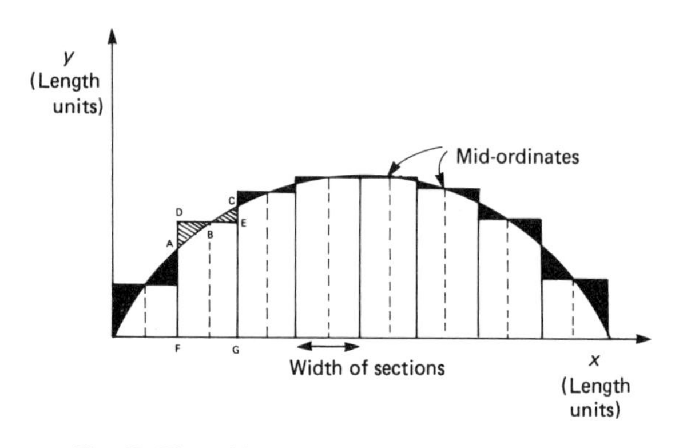

Fig. 49. *The mid-ordinate rule for area calculation*

the midpoint of its base to the curve. Such lines are dashed in Fig. 49.

The perpendicular distance of a point to the x-axis is known as its *ordinate*. As the dashed lines in Fig. 49 not only represent ordinates, but also pass through the midpoint of the base of each rectangle, they are known as *mid-ordinates*. This is how this method of finding the area beneath a curve gets its name.

By calculating the product of the average length of the mid-ordinates, the width of one section and the numbers of sections we can approximately find the area beneath the curve. Obviously, the larger the number of sections we take, the more accurate will be the result.

Basic Trigonometry

Let us now revise the simple principle of trigonometry, in which we deduce the sine of an angle. For angles less than 90° the sine of an angle is simply the ratio of two sides of a right-angled triangle. These sides are given names as shown in Fig. 50. When x is being considered, the side opposite to x is called the *opposite side* BC. The side AC is called the *adjacent side*, being next to the angle. The side opposite the right angle, is always known as the *hypotenuse*.

The sine of x is expressed as the ratio:

$$\sin x = \frac{\text{opposite}}{\text{hypotenuse}} = \frac{BC}{AB}$$

When we know the length of AB and we know the angle x, the length of BC can be calculated since BC must $= AB \times \sin x$. It

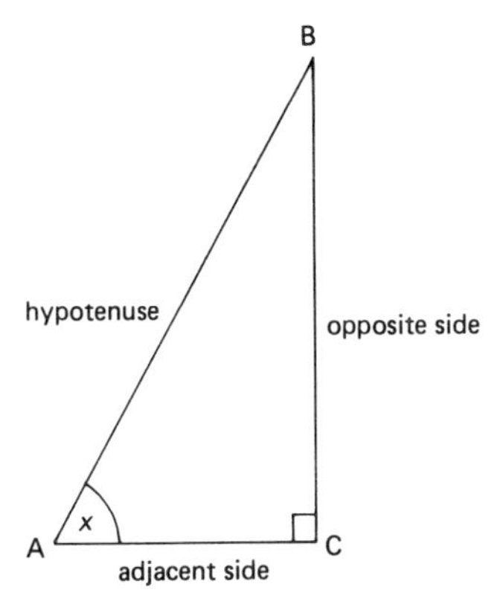

Fig. 50. *Basic trigonometrical terms*

should be noted that for angles greater than 90° the value of sine can be found by observing the symmetry of a sine wave, taking into account whether the value of the sine is positive or negative.

ALTERNATING AND UNIDIRECTIONAL WAVEFORMS

The electric current consists of a movement of charged particles along a physical conductor. When a certain number of particles move, a current of a particular value is said to flow, and this can be measured. This can be represented on a graph as in Fig. 51(*a*).

When the current value changes, let us say that it is halved, then Fig. 51(*b*) would represent this change. Similarly, every other change can be reflected in the shape of the curve. We must note carefully that in this case the charged particles, which we know are electrons, always continue to move in the same direction, even though more or fewer will move depending upon the value of the current flowing. This unidirectional flow is called a direct current and the electrons will never move in the opposite direction, provided the source of e.m.f. in the circuit does not change polarity. Certain engineering advantages can be gained by the application of an e.m.f. with reversing or alternating polarity. A rotating generator naturally generates an alternating voltage, and when this is applied to a circuit the electrons are stimulated to flow, first in one direction and then in the other. A graph can also be drawn representing this, but clearly some method of distinguishing between the directions of current flow

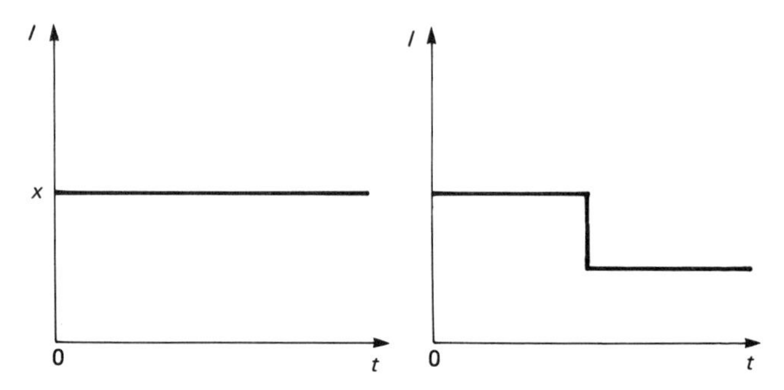

Fig. 51. (a) *Constant d.c. current;* (b) *d.c. current halved*

has to be found. The mathematician has a simple way. He chooses to call current in one direction *positive*, and in the other direction *negative.* Using this simple method, we can represent both directions on the same graph without difficulty. An alternating current is shown in Fig. 52(*a*). For comparison a varying direct current is shown in Fig. 52(*b*).

The actual shape shown here is quite complicated and it is best to consider first the regular shape which we can assume is generated by the simple rotating generator. Figure 53 shows a simple coil rotating between the poles of a magnet and generating the alternating shape shown in Fig. 54. It is known as a *sinusoid.* The graph itself is known as a sine wave form. There are several important new terms which must be learned to describe this alternating wave form and these are shown in Fig. 55.

The maximum magnitude that the current or e.m.f. attains is termed the *amplitude*, or *peak value.*

The *peak to peak value* is that measured between the peaks as shown, i.e. from maximum positive to maximum negative values of current or e.m.f.

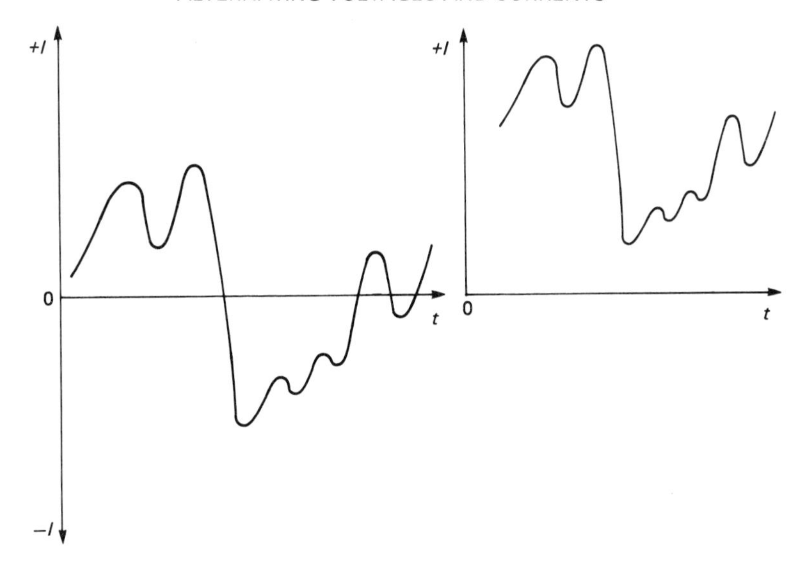

Fig. 52. (a) *Alternating current;* (b) *fluctuating direct current*

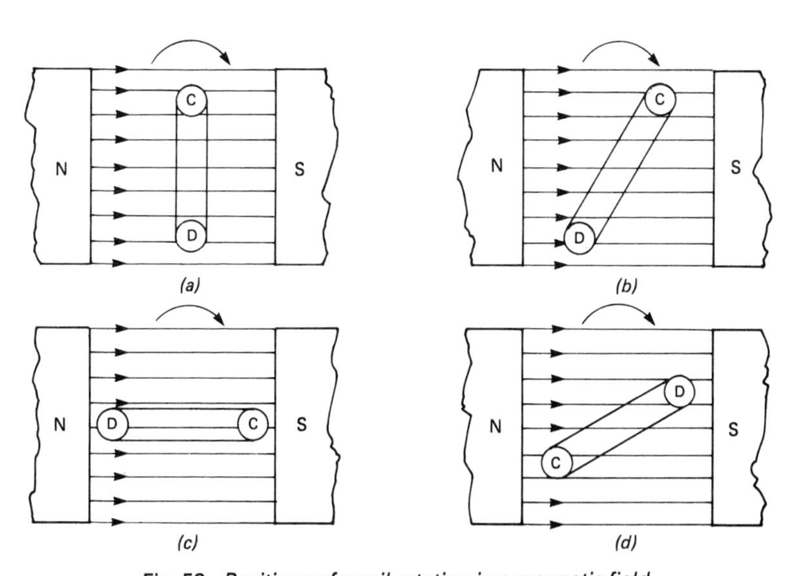

Fig. 53. *Positions of a coil rotating in a magnetic field*

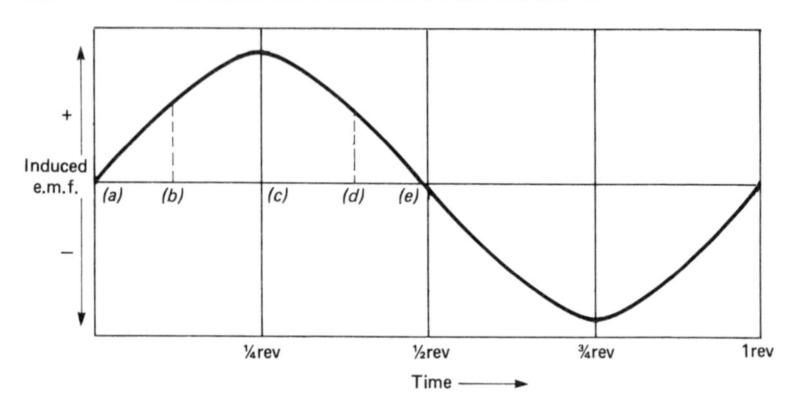

Fig. 54. *Sine wave generated by the coil in Fig. 53*

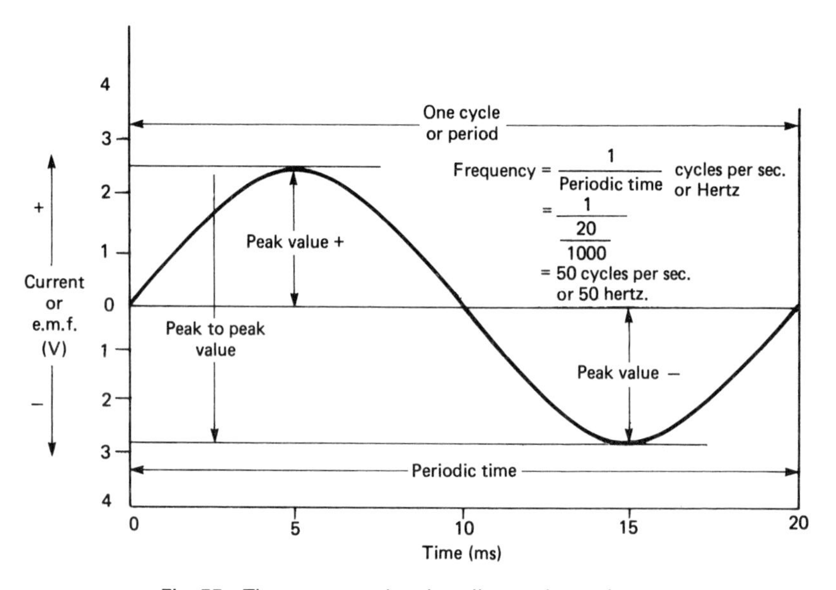

Fig. 55. *The terms used to describe an alternating wave*

The *frequency* of a repetitive wave form is the number of completed cycles per second. The diagram shows one complete cycle. The total time taken for one cycle is 20 ms, hence the frequency would be 50 cycles per second or 50 Hz. In practice, 50 Hz is a low value and is the mains frequency.

The time it takes to complete one cycle is known as the *period* or *periodic time* and is 1/frequency, seconds, e.g. what is the periodic time of an alternating voltage of frequency 100 Hz? Periodic time = 1/100 = 0.01 s.

INSTANTANEOUS, AVERAGE AND R.M.S. VALUES

Instantaneous Value

We know already that it is the flow of current which gives many useful effects which we can use to our advantage. For example, the amount of heat generated in a circuit will depend upon the amount of current flowing. In the case of the constant value direct current, this presents no difficulty. The value x from Fig. 51(a) can be used in any calculation to find the amount of heat generated.

The alternating wave form, however, presents problems and other values of importance must be considered. The current flow is constantly changing and this must be taken into account. One measurement which may be needed is the value of the current or voltage at given instant. This is appropriately known as the *instantaneous value*. It can lie anywhere between the positive and negative peak values, and will vary from instant to instant. It must be realised that it does not matter in which direction the electrons are flowing, the various effects due to the current will still be taking place.

Average Value

At some instants during the cycle, the value of the current will be zero. It would therefore seem reasonable to determine an average value, of current flow. The mathematical average taken over a *full* cycle of a sine wave would, obviously, be zero. However, it is acceptable to take the average value of a half sine wave since, as explained, the negative half cycle is only another

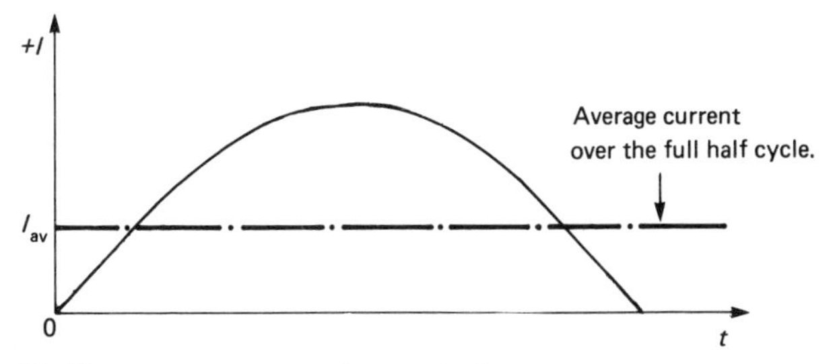

Fig. 56. *The current averaged over a half cycle of an alternating wave*

way of saying that the current reverses its direction. This result gives the average magnitude of the current. This is the value shown as a dotted line in Fig. 56. The average for a sine wave can be shown to be 0.637 × peak value or $2/\pi$ × the peak value attained. Therefore, if it is known that the waveform is a true sinusoid, a simple calculation can be made to find the average magnitude of the current.

R.M.S. Values

A second important value, but one which at first seems to have an unusual derivation, is the *root—mean—square* or *r.m.s.* value. It means exactly what it says, and it will be best appreciated by taking the derivation step by step.

In earlier work we have seen that the heating effect of a direct current can be calculated by using the simple expression I^2R, where I is the current and R is the resistance of the circuit through which it flows. In the case of an alternating current, each instantaneous value of the current will contribute a corresponding amount of heat. It would seem logical, therefore, to square every instantaneous value of I and then calculate the mean value of I^2 (the mean-square value) which we will call Q. Thus, the mean heat generated will be QR.

It is useful to determine the direct current which will produce the same mean heating effect as the above alternating current (if both are passed through the same circuit). Let us call this direct current I_{dc}. Thus:

$$I_{dc}^2 \times R = Q \times R$$

$$\therefore \quad I_{dc}^2 = Q$$

$$\therefore \quad I_{dc} = \sqrt{Q} \ .$$

The current I_{dc} is therefore equal to the square root of Q, the mean-square value of the alternating current, and is called the root-mean-square value.

The r.m.s. value is very important. It effectively tells us what work, e.g. heating, an alternating current can do for us.

The r.m.s. value of a sine wave is found to be 0.707 × I_{max}. When other than sine waves are being considered, this value will change and a comparison of amount of change can be used to indicate the overall shape of a given wave form.

The actual r.m.s. and average values of non-sinusoidal wave forms, can be calculated by taking the root of the mean of the squares, or using the mid-ordinate rule respectively.

Form Factor

When sine waves are being considered, the ratio of average to r.m.s. values is calculated thus:

$$\text{ratio} = \frac{I_{\text{r.m.s.}}}{I_{\text{av}}} = \frac{0.707 \times I_{\text{max}}}{0.637 \times I_{\text{max}}} = \frac{0.707}{0.637} = 1.11.$$

This ratio is given the specific term *form factor.* Thence by definition, form factor is the ratio of r.m.s. value to average value for the wave form. The power engineer uses the form factor quite extensively.

When a generating plant operates, it tends to introduce voltage harmonics, i.e. multiples of the basic generated frequency. These harmonics may cause trouble in the circuits to which they are fed. When the wave shape varies from the ideal sinusoidal shape this indicates the introduction of unwanted frequencies. The average and r.m.s. values will also vary and hence the form factor will change. The amount which the form factor deviates from 1.11 is a good indication to the Engineer about the shape of the generated wave form. A lower value indicates a *squarer than sinusoidal wave*, whereas a higher value indicates a more *triangular shape.*

PHASOR AND ALGEBRAIC REPRESENTATION OF ALTERNATING QUANTITIES

Phasor Quantities

A vector quantity must have two qualities to describe it fully. It must have size and direction. Let us say that an astronaut wishes to know about the qualities of his rocket motor. He needs to be aware how much thrust it will generate, so that he knows the rocket will launch. Equally important is the fact that he needs to know that the thrust will be upwards and not sideways, or the rocket would tip over.

It is quite acceptable to say that the motor will develop 200 t thrust vertically down towards the surface of the Earth. This will produce a force on the spacecraft of 200 t vertically upwards from the surface of the Earth. An alternative and perhaps more simple way of describing this would be to draw a diagram and represent the force on the space craft as shown in the Fig. 57(*a*).

The line drawn **A**, represents both the force and its direction. However, the rocket has a physical weight due to the gravitational pull. Let us say that it is 175 t. We know the gravitational pull is vertically downwards and this can also be represented by a vector **B**. This is shown in Fig. 57(*b*).

It does not require too much imagination to realise that the two diagrams may be usefully combined to tell the full story (*see* Fig. 57(*c*)).

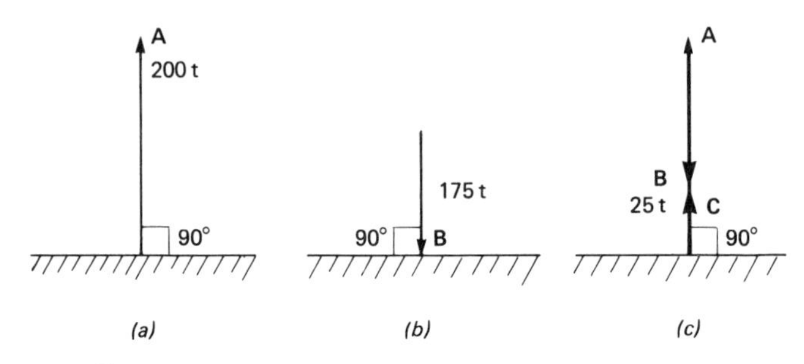

Fig. 57. *The principle of using vectors to find a resultant*

This shows that the resultant force is 25 t vertically upwards, as represented by **C**.

This simple example illustrates a very important principle, that a diagrammatical representation of vectors can be achieved showing size and direction, and that the resultant effect may be deduced by using it. The simple diagram is called a *vector* diagram.

The Phasor

Imagine now that the size of the force which we wish to represent does not remain constant, but varies with time. Provided the variation follows a pattern which repeats on time, this can also be represented by a single line. However, it is then assumed that the line will rotate with a certain angular velocity or rotating speed, to trace out the variation in full. Consider the diagram shown in Fig. 58. The line AC, which we assume starts horizontally at "3 o'clock", rotates anticlockwise. After it has traced out 30° the length of AB will have reached a certain value.

A few degrees later it will be larger, and after it has traced out an angle of 90° it can be seen that it will be a maximum, and as its angle to its original position increases it will start to reduce again. Draw a sketch for yourself and trace out the full pattern

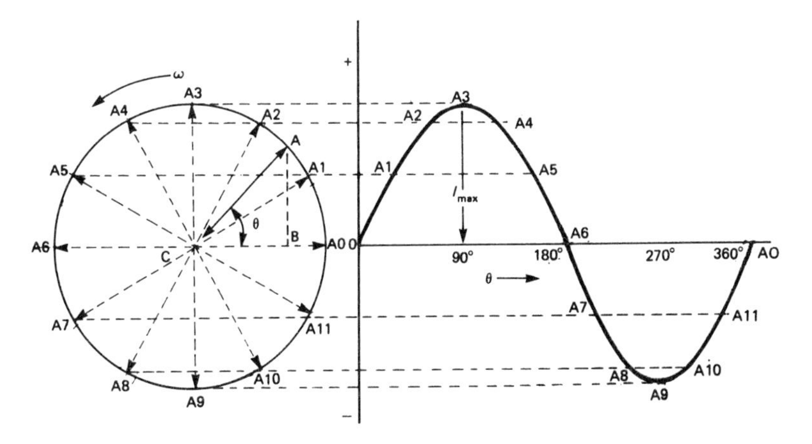

Fig. 58. *A rotating phasor tracing out a full sine waveform*

followed by the values of AB. Assuming that AB rotates at a constant rate, it is not too difficult to realise that a graph can be drawn showing the changing values of AB with a horizontal time axis.

Figure 58 shows this pattern. A careful comparison with Fig. 55 earlier in this chapter, shows that the pattern traced out is a sine wave. The length of AC represents the amplitude of the sine wave. It seems reasonable then to say that a single line can be made to represent the values of a sine wave, provided that it is assumed to rotate. This is so, and it is often done. To distinguish it from the vector discussed earlier, it is called a phasor, i.e. a rotating vector. It must now be appreciated that all the values traced out are the instantaneous values of the sine wave at these points of time.

You may well be asking what the point is. In just the same way as the vector diagram was a convenient way of representing the force on the spacecraft so that the resultant force could be determined, the following example shows how the phasor diagram can help.

Assume there are two alternating voltages applied to a circuit. They have the same frequency, but different amplitudes. We wish to know what the total voltage at any instant will be. One method is to draw the two sine waves carefully on the same axes, and sum them point by point (*see* Fig. 59).

By adding A to A_1, B to B_1, C to C_1, etc. we will produce the resultant voltages shown at D, E and F. Let us now investigate a

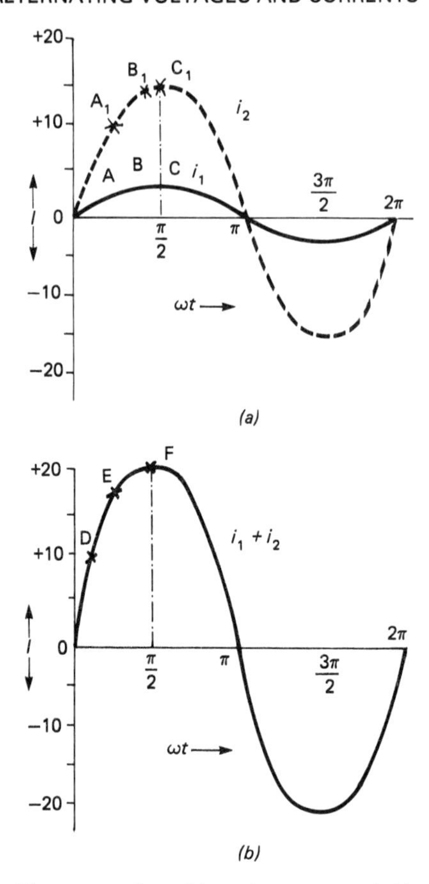

Fig. 59. *The summation of two sine waves point by point*

simpler method. We know the value at C is the maximum for one wave and that at C_1 the maximum for the other. Now, draw two phasors representing the two sine waves (*see* Fig. 60). Add the two phasors together and the resultant is shown as $C + C_1$. When

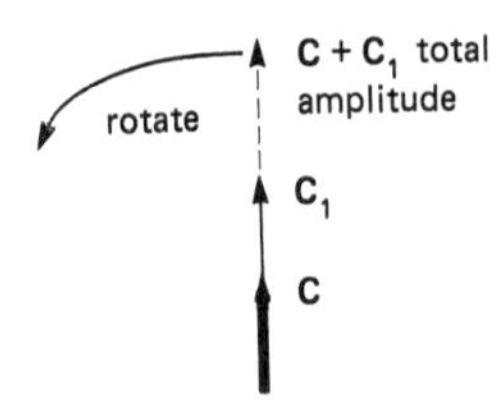

Fig. 60. *The phasor method of summing two sine waves*

this resultant phasor is rotated, it will trace out the same curve as that shown in DEF, and the complex problem of adding the graphs together is avoided. This principle can be applied for all sinusoidal quantities and is a useful tool for the electrical engineer.

Algebriac Representation of a Sine Wave

Referring back to the simple trigonometry at the start of this section, you will remember that the length of AB can be calculated from AB = AC sin θ but we already know from Fig. 58 that AB represents the instantaneous values of the sinusoidal current or voltage we are considering. Also we know that AC will be the maximum value of the phasor and the angle θ will be the angle traced out by the rotating phasor. This then becomes a simple algebraic way of representing the sine wave.

$$v = V_{max} \sin \theta$$

Where v = instantaneous value

V_{max} = maximum value

θ = angle traced out in degrees.

We have already discussed that the angle can be measured in radians. When this is the case, the angle traced out will depend upon how long an interval of time has passed and how fast the phasor is rotating. Hence, the angle can be deduced from the product of the speed of rotation in radians/second and the time elapsed in seconds. This may be expressed thus:

ω = angular velocity rad/s.

$$v = V_{max} \sin \omega t.$$

t = time in seconds.

The phasor and algebraic methods are both important ways of describing the sine wave and these methods must be learned.

Phase Relationships between Two Rotating Phasors

It is easy to imagine that when two voltage wave forms reach their maximum value at the same time, the total instantaneous voltage is the two added together. However, it is a much more difficult problem when the maxima are not at the same instant. Figure 61 shows two voltages of the same frequency but with one peak behind the other in time. The resultant instantaneous voltage is obviously much more complex. When two wave forms have this off-set time relationship, as have those shown, they are referred to as having a *phase difference*. The phase difference may be measured as an angle between them. In the figure it can

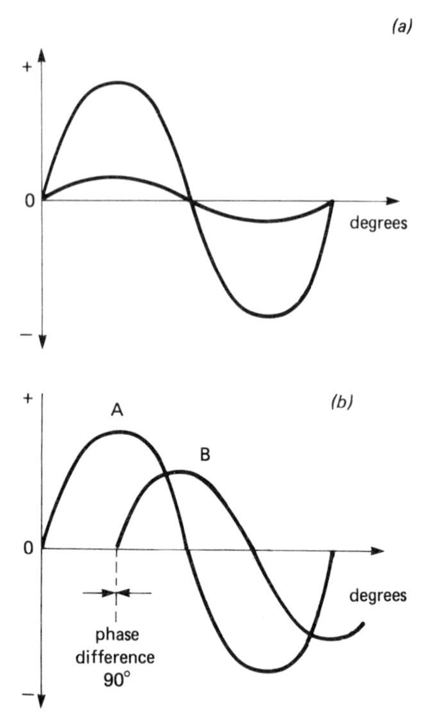

Fig. 61. *An example of waveforms:* (a) *in phase; and* (b) *out of phase by 90°*

be seen that the B wave is behind. This is commonly known as B lagging A by a quarter cycle, i.e. by 90° or $\pi/2$ radians, depending upon which angular measure we are using. Alternatively A may be described as leading B by $\pi/2$ radians.

Phase differences are met in many different forms of electrical engineering and many side effects result from it.

It may be necessary to describe the wave form using a mathematical expression, but at the same time show that a phase difference exists. It would seem quite reasonable that the phase difference of so many radians is simply added on, or subtracted from, the expression for the original wave. It is perfectly logical therefore to express this new wave as:

$$v = V_{max} \sin (\omega t \pm \varphi)$$

where φ represents the phase difference.

Since φ is in the form of an angle it is often referred to as the

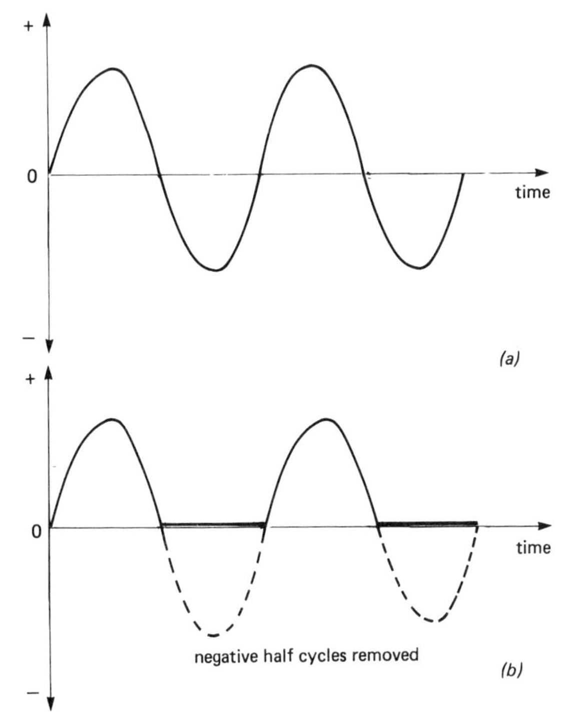

Fig. 62. *Rectification showing:* (a) *original a.c. wave; and* (b) *rectified wave*

phase angle. The wave B, shown in Fig. 61, could therefore be expressed as:

$$v = V_{max} \sin\left(\omega t - \frac{\pi}{2}\right) V.$$

THE PRINCIPLE OF RECTIFICATION

Earlier in this chapter we defined alternating and direct voltages and currents. It was stated then, that any current which did not reverse its direction could be described as a direct current. An alternating current was said to be one which changed its direction regularly (for instance, a sine wave varies with a continuous pattern).

Looking at the sine wave again, the most obvious way to convert from the alternating to the direct form is to cut off all of the negative part of the graph (*see* Fig. 62).

The practical application of this simple principle is known as rectification. Naturally, the "lumpy d.c." which remains is not too useful without further treatment, but rectification is the first step.

How do we do this cutting off the negative half cycles practically? Imagine that the waveform is alternating very slowly and that a switch has been included in the circuit as shown in Fig. 63.

The circuit also has an indicator which shows the generated voltage. The operator carefully observes the generated voltage, and closes the switch only during the positive half cycles. As far as the load is concerned, which in our simple diagram is a

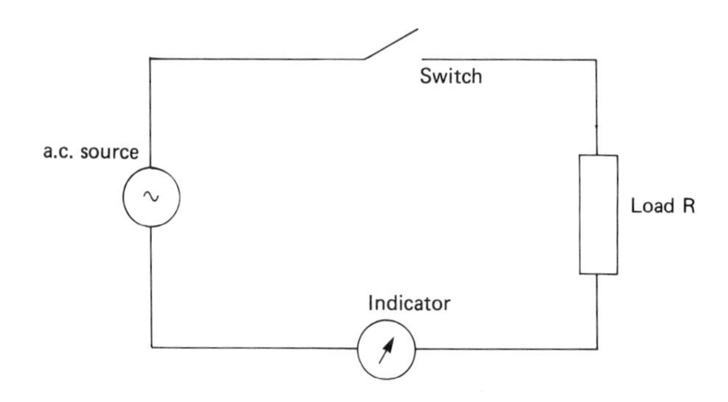

Fig. 63. *The simple principle of rectification*

resistor, it only has direct current flowing through it. This is a very important principle. The generator is still generating an alternating voltage which, without interference, would drive an alternating current through the load. It is the switch which modifies the circuit to allow only direct current to flow through the load.

In practice, the switching process must obviously be automatic, but what happens is identical. Any device which allows a current to flow through itself in one direction only, will do the job of an automatic switch. A device which is commonly employed is a semi-conductor junction diode. This has excellent characteristics for the job, offering little forward resistance but very high resistance in the reverse direction.

HALF-WAVE AND FULL-WAVE RECTIFICATION

The rectification process just described is very wasteful. The negative half cycle, although generated, is put to no useful purpose. This process is known as half-wave rectification. A simple half-wave rectifier circuit is shown in Fig. 64. Figures 64(*b*) and 64(*c*) show the effective operating condition during the two half-cycles and Fig. 65 shows the waveforms of the generated current and the load current.

It is well within the ingenuity of engineers to work out a circuit which would make use of the wasted half-cycle. Consequently, the circuit produced to do the task is known as a full-wave

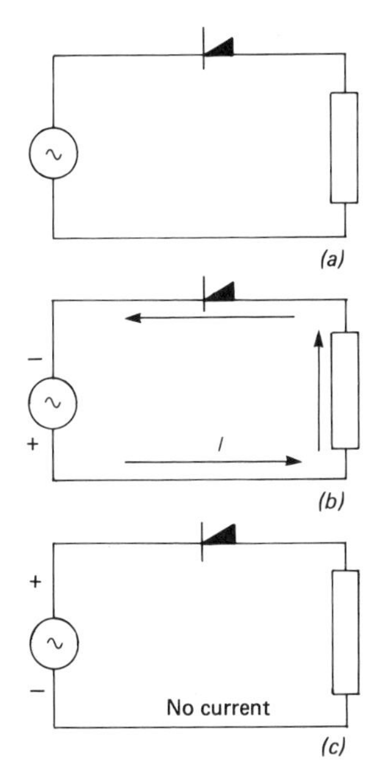

Fig. 64. *A simple half-wave rectifier circuit and its operation over each half cycle*

rectifier. Its design is such that the negative half-cycle, instead of being wasted, is driven through the load in the same direction as the positive half-cycle. The load current waveform and the generated waveform are shown in Fig. 66.

A simple full wave rectifier circuit is shown in Fig. 67. The current directions are shown in full line and dotted line for the positive and negative half cycles respectively.

It has already been mentioned that the resultant d.c. shown in Figs. 65(*b*) and 66(*b*) are very poor for practical uses. However, simple techniques are available to help improve the quality of the

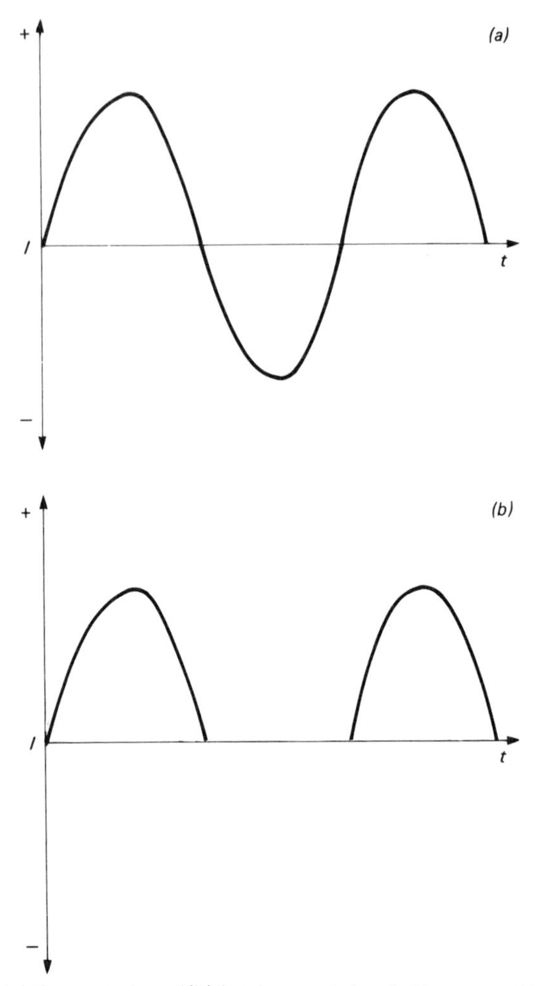

Fig. 65. (a) *Generated; and* (b) *load currents in a half-wave rectifier circuit*

d.c. The process is known as *smoothing*. The subject of smoothing is outside the scope of this book but a simple circuit is given to show a typical method (*see* Fig. 68).

THE RELATIONSHIPS BETWEEN GRAPHICAL, PHASOR AND ALGEBRAIC REPRESENTATION OF SINE WAVES

So far we have carefully built up three different methods of expressing a sinusoidal voltage or current. We must now be able

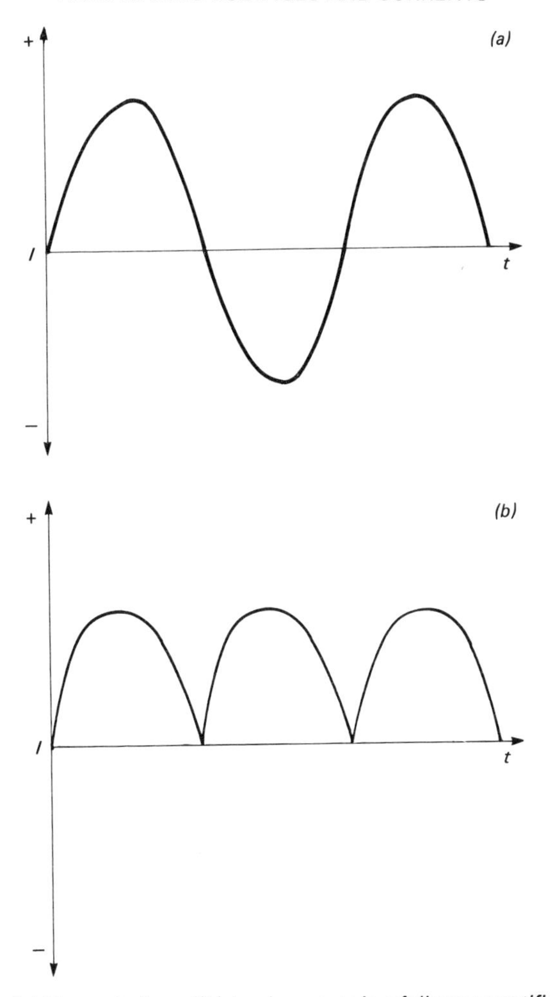

Fig. 66. (a) *Generated; and* (b) *load currents in a full-wave rectifier circuit*

to relate each method to the others, so that a full understanding of the methods is achieved.

The relationship between the three must be considered for each of the main aspects of the sine wave:

(*a*) amplitude;
(*b*) frequency; and
(*c*) phase.

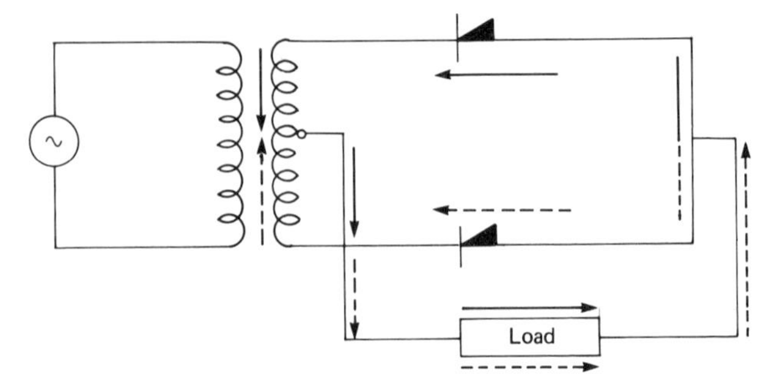

Fig. 67. *A full-wave rectifier circuit*

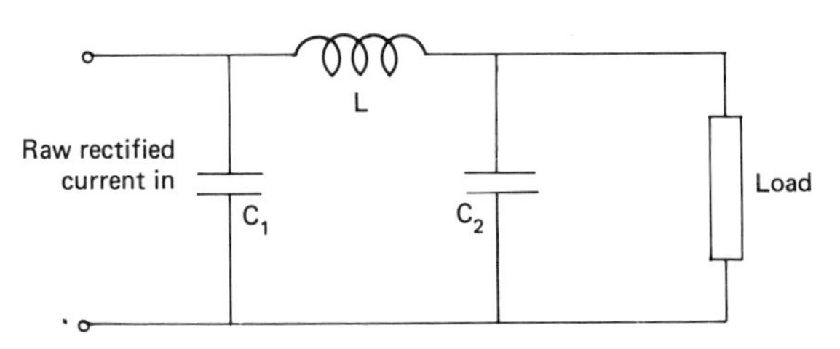

Fig. 68. *A simple smoothing circuit*

Amplitude

The peak, or maximum value, of the sine wave is the maximum reading achieved on the graph. This same value is represented by the length of the phasor and is represented by E_{max} or I_{max} in the algebraic expression.

Frequency

The frequency can be deduced from the graph by observing the time scale and extracting the time taken to complete one cycle. The frequency can then be calculated.

$$\text{Frequency} = \frac{1}{\text{periodic time in seconds}} \text{ Hz}$$

The phasor diagram should show the angular velocity of the rotating phasor in rads/s. It is assumed that the phasor rotates in an anticlockwise direction represented as in Fig. 60.

As we have already considered, there are 2π radians in 360°, hence there must be $2\pi \times$ frequency radians/second being traced out. The frequency is therefore deduced from ω the angular velocity,

$$\text{and} \quad f = \frac{\omega}{2\pi} \text{ Hz.}$$

We know that the algebraic expression also contains the ω term, i.e:

$$[v = V_{max} \sin(\omega t + \varphi)]$$

and the frequency can again be deduced from ω as above.

Phase

We have discovered that phase is also important and each method must be capable of telling the reader how the phase relationship can be determined.

Referring back to Fig. 61, it can be seen that the deviation between the maxima or zeros can be read from the scale directly in terms of a phase angle. It may be expressed in radians or degrees and does not vary for the particular wave form.

The phase angle between two phasors is easily represented, simply being shown as an angle between the two. Figure 69 shows

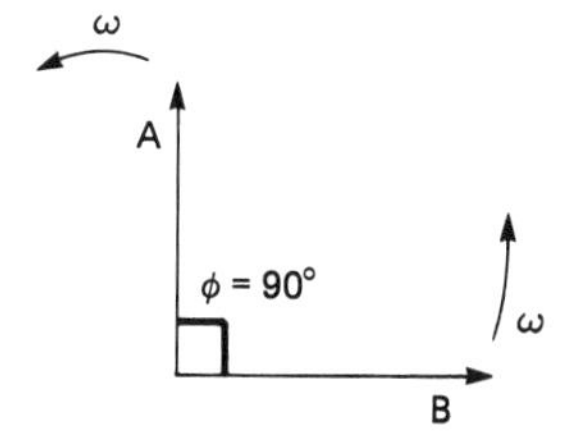

Fig. 69. *Phasor representation of the two waveforms shown in Fig. 61*

the phasor representation of the two waveforms expressed graphically in Fig. 61.

The algebraic expression shows up any phase deviation relative to the original wave directly. As such the wave form B is shown as:

$$v = V_{max} \sin\left(\omega t - \frac{\pi}{2}\right).$$

SELF-ASSESSMENT QUESTIONS

1. When a loop of wire is rotated at a constant speed and at right-angles to a magnetic field, the voltage induced across the ends of the loop is a:
   (*a*) direct voltage;
   (*b*) fluctuating direct voltage;

(c) sinusoidal alternating voltage;
(d) irregularly alternating voltage.

2. Calculate the r.m.s. value of an alternating voltage having the following values at equal time intervals, the points being connected by straight lines: 0, 5, 10, 20, 50, 60, 50, 20, 10, 5, 0, −5, −10, −20, −50, −60, −50, −20, −10, −5 and −0. What would be the r.m.s. value of a sinusoidal wave having the same peak value?

3. A sinusoidal current has a r.m.s. value of 6A and a frequency of 50 Hz. Determine the time taken for the current to increase from zero to 5A and the value of current 2.3 ms after passing through zero, increasing in a positive direction.

4. Draw phasors, on the same diagram, to represent the following currents:

$$I_1 = 6 \sin \omega t;\ I_2 = 10 \sin (\omega t + \pi/4);\ \text{and}\ I_3 = 4 \sin (\omega t - \pi/3)$$

Construct the resultant phasor and measure the resultant and its phase relationship with I_1, express the resultant in the form $I = I_{max} \sin (\omega t \pm \varphi)$.

5. Draw the waveform of a simple sinusoidally varying voltage and label on it:
(a) the peak to peak voltage;
(b) the periodic time;
(c) one complete cycle;
(d) the maximum voltage.

6. Describe the three possible methods of representing a sinusoidally alternating voltage giving examples of all three.

7. From the diagram given:
(a) state the phase relationship between A and B;
(b) say which waveform has the highest frequency;
(c) draw a phasor diagram representing the given waveforms.

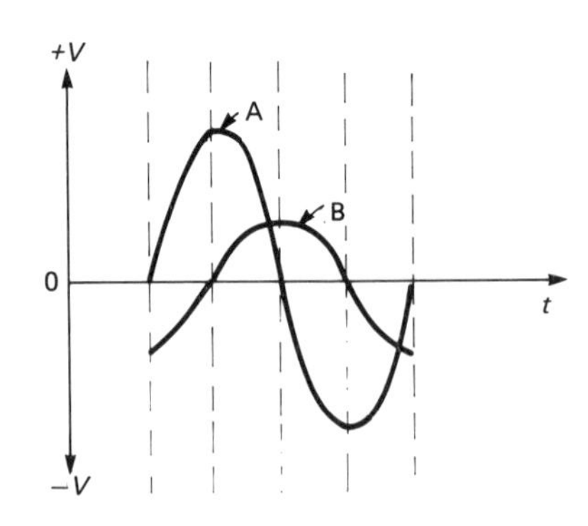

8. A current is given as $I = I_{max} \sin(2\pi ft + \varphi)$.
Describe carefully what each symbol represents.

9. A voltage is given as $V = 100 \sin 800\,\pi\,t$.
Deduce:
(*a*) the maximum voltage;
(*b*) the r.m.s. value of voltage;
(*c*) the frequency of the waveform;
(*d*) the periodic time.

10. Describe what the power engineer refers to as the *form factor*.

11. Describe the basic principle of rectification and describe a simple full-wave rectifier circuit.

CHAPTER SIX

Single Phase a.c. Circuits

CHAPTER OBJECTIVES

After studying this chapter you should be able to:

* understand the phase relationships between voltages and currents in all cases where reactive components form part of a circuit;
* understand the relationship between reactance and impedance;
* represent the circuit operating conditions for currents, voltages, power and series resonant conditions with the aid of phasor diagrams.

INTRODUCTION

When considering the behaviour of single phase a.c. circuits it is important to realise that instantaneous values of current and voltage are of little use in circuit calculations. It is much of more use practically to deal with the r.m.s. values, as was explained in Chapter 5. It is therefore common practice to assume that all quoted a.c. values are r.m.s. values unless otherwise stated, and these will be used in the circuit calculations in this section.

In all d.c. circuit calculations it is assumed that the component values always remain the same ignoring any minor changes of resistance due to temperature. However, this cannot be assumed when alternating voltages and currents are being considered. Furthermore, it has always been assumed that a change in applied direct voltage gives rise to an instantaneous change in current if the short term effects of inductance are not considered. This again is subject to doubt in the case of alternating circuits and the phase relationship between the voltage applied across a component and the current through it must be carefully examined. In fact the three main circuit components, i.e. resistance, inductance and capacitance must all be examined separately in order to understand their circuit behaviour.

A PURE RESISTANCE IN AN a.c. SINGLE PHASE CIRCUIT

The simplest circuit configuration for a pure resistor and an a.c. source is shown in Fig. 70.

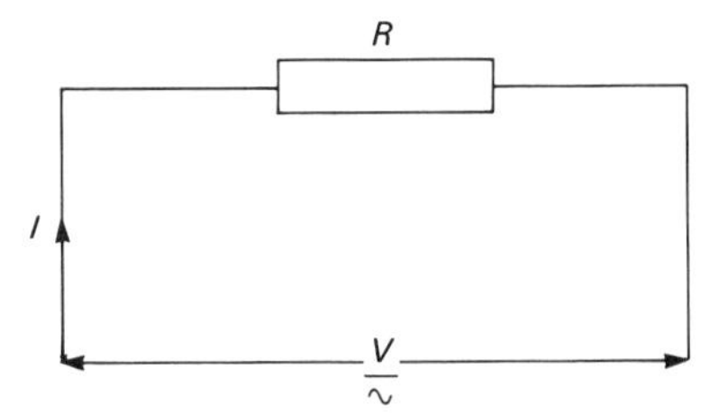

Fig. 70. *A.c. supply connected to a pure resistance*

When the voltage V is applied the following facts can be deduced.

(*a*) The value of current flowing is given by

$$I = \frac{V_r}{R} \text{ amperes}$$

where V_r is the applied voltage across R, and R is the resistance in ohms.

(*b*) When the frequency is changed and the r.m.s. voltage remains constant, the value of r.m.s. current remains constant. It can therefore be deduced that the value of R does not vary with frequency.

(*c*) The phase relationship between the applied voltage and the current produced is that they are in phase, i.e. they reach their maximum, minimum and zero values at the same time. This phase relationship can readily be shown using a phasor diagram Fig. 71.

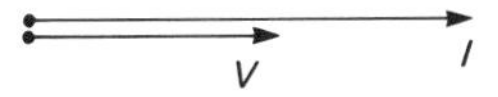

Fig. 71. *Phasor diagram showing voltage and current in phase in a purely resistive circuit*

The characteristics of the resistance in an a.c. circuit can thus be summarised by saying that it behaves exactly as it would in a d.c. circuit.

A PURE INDUCTANCE IN A SIMPLE a.c. CIRCUIT

When a pure inductance is connected to an alternating voltage (*see* Fig. 72), a current will flow through it. However, in this case it displays properties which have not been met with before in d.c.

circuits. Firstly, if the frequency of the applied alternating voltage is changed, we find that the value of current being driven round the circuit also changes. We have already observed that an Ohm's law type of relationship exists in the a.c. circuit, hence, if the current has changed whilst the voltage remains constant then the opposition must also have changed. It is found that as the

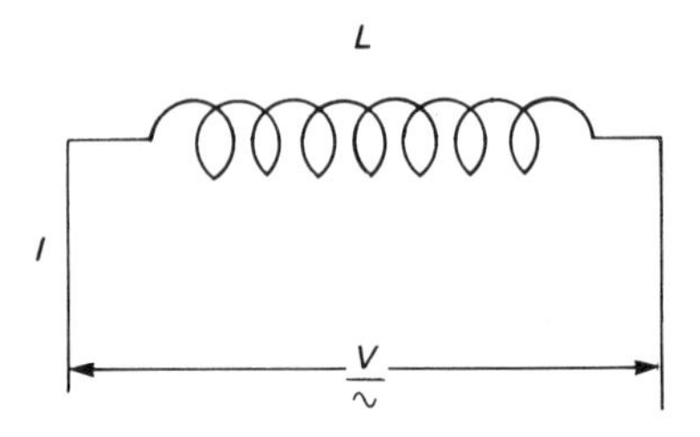

Fig. 72. *A.c. supply connected to a pure inductance*

frequency increases, the current will fall. The opposition to the current must therefore have increased. Indeed this is the case and we will consider the exact relationship in due course.

Inductive Reactance

It seems that we shall have to introduce a separate term for this changing opposition. We know that the value of the resistance does not change with changing frequency, therefore it cannot be labelled resistance. The word chosen is reactance, and since the reactance is associated with an inductance the full term used is the *inductive reactance.*

The Voltage/Current Phase Relationship

In an a.c. circuit we have already introduced the idea that phase as well as size is important. We must now look at the voltage and current phase relationship when the voltage across and the current through a pure inductance are being considered. It is found that the two differ considerably in phase, their peaks and zeros being separated by a full 90°. This is best shown in Fig. 73.

The voltage always leads by 90° in a pure inductance. This is a fact which must be learned, and applies no matter what the frequency of the supply voltage may be.

The Phasor Diagram for a Pure Inductance in an a.c. Circuit

We can use a phasor diagram to show this out of phase condition very easily provided some basic rules are remembered:

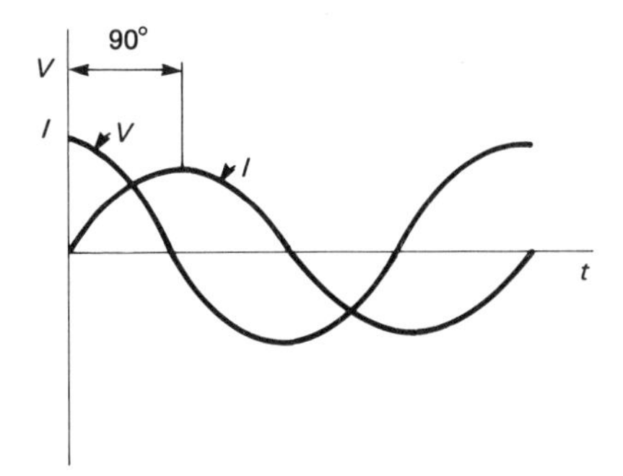

Fig. 73. *Phase relationship of current and voltage in a purely inductive circuit*

(*a*) that the reference is always drawn at "3 o'clock"; and
(*b*) that leading phasors are always assumed to be displaced anticlockwise.

In Fig. 72 the current is common, passing through each part of the circuit, and can therefore be used as a reference. The phasor diagram is therefore as shown in Fig. 74.

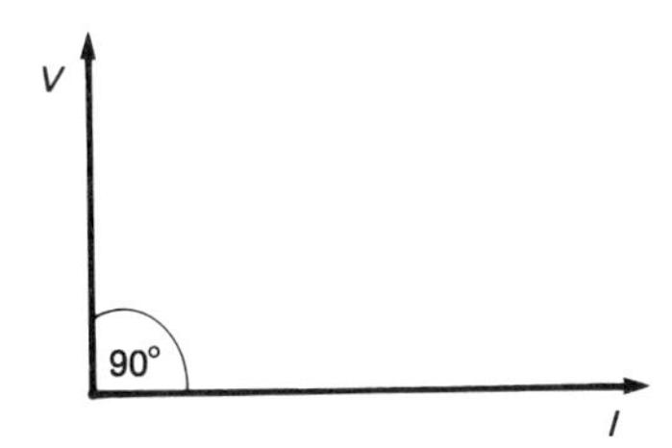

Fig. 74. *Phasor representation of the phase relationship between voltage and current in a purely inductive circuit*

It should be noted that the phasor diagrams are representative of circuit conditions and are not drawn to scale.

A PURE CAPACITANCE IN A SIMPLE a.c. CIRCUIT

When an alternating voltage is applied to a pure capacitance a current will be driven through it. The capacitance will therefore allow an alternating current to flow through it. Let us again assume that the frequency of the supply is changed. In this case we shall find that the current will increase as the supply frequency increases. The opposition to current flow must therefore be reduced. Again, we shall consider the amount of the change a little later.

Capacitive Reactance

One again, it is obvious that the opposition does not behave as a resistance and to call it resistance is unsuitable. The term reactance is used once more but on this occasion to distinguish it from that of the inductance the full term *capacitive reactance* is used.

The Voltage/Current Phase Relationship

A careful look at the changing voltage and current waveforms will show that there is a phase difference of 90° in this case also. There is a major difference however, in that the 90° is in the opposite sense, and the current flowing in a capacitor leads the voltage across it. The waveforms are shown in Fig. 75.

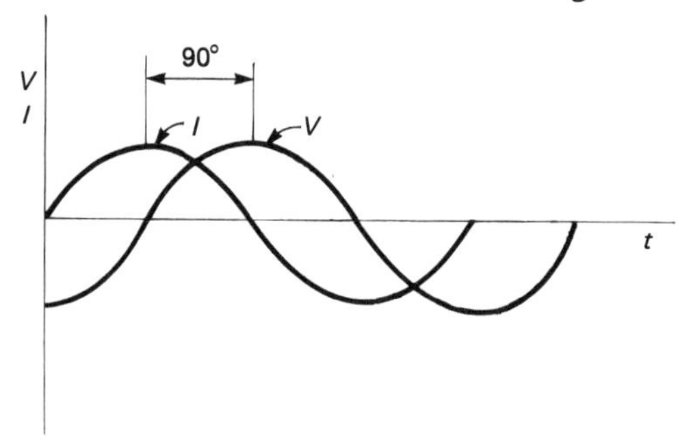

Fig. 75. *Graphical representation of the phase relationship between voltage and current in a purely capacitive circuit*

The Phasor Diagram for the Pure Capacitance in an a.c. Circuit

Using the same rules as previously the phasor representing the current I and the voltage V_c can be drawn in Fig. 76.

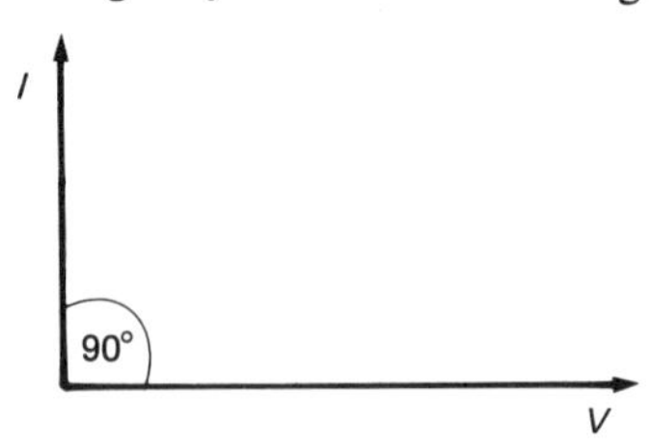

Fig. 76. *Phasor representation of the phase relationship between voltage and current in a purely capacitive circuit*

POWER DISSIPATED IN A PURELY INDUCTIVE OR CAPACITIVE CIRCUIT

The power in the d.c. circuit was shown to be a product of volts and amperes. The power used being measured in *watts*. The volts and amperes product can also be used to find the a.c. power being used, but it must be remembered that these values are constantly changing. Let us divide up Fig. 77 into four separate

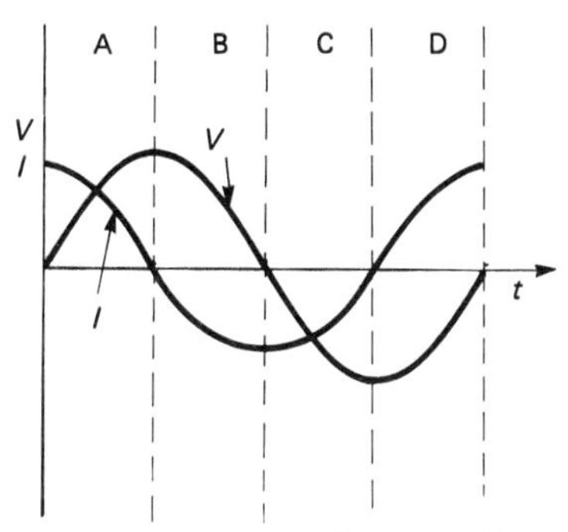

Fig. 77. *Voltage and current waves in a capacitive reactive circuit*

sections A, B, C and D. If we consider each section in turn, the product of voltage and current can be taken, paying special attention to whether or not the voltage and current are positive or negative. Look at section A. Both are positive and hence the product will also be positive. Section B however, has the current negative and the voltage positive, giving a negative result. In section C both are negative and the product becomes positive once more, whereas in section D a positive current is counteracted by a negative voltage. A power graph can thus be deduced and drawn. The result is presented in Fig. 78. A similar power

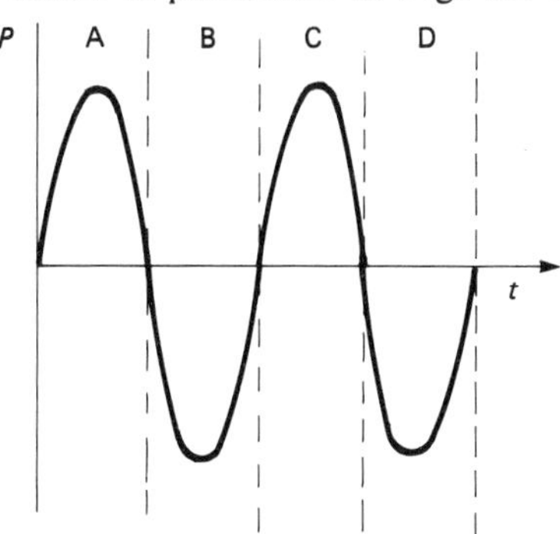

Fig. 78. *Power wave for the capacitive reactive circuit*

graph can also be drawn for the pure inductance case, using exactly the same reasoning.

The Average Power

It is not too difficult to observe that the average power dissipated, taken over the full cycle is zero. This is demonstrated graphically in the Fig. 78. This may seem difficult to accept, but it is the case. Careful experiments have shown that purely reactive components do not dissipate any power at all. The energy is fed to the component during one part of the cycle, stored in the inductance or capacitance as a magnetic or electrostatic field, and then returned to the source during a later part of the cycle. In practice, a purely reactive component cannot be produced and power is dissipated in the physical resistance of the inductor or capacitor. Nevertheless, capacitors use up only a little power and a well designed coil may have only a low loss in practice. Both are very useful components particularly to the electronic and communications engineer.

DEPENDENCE OF INDUCTIVE REACTANCE ON FREQUENCY

Let us return to the variation in opposition to the flow of an alternating current. Again, from laboratory experiments it is found that when the frequency is changed the inductive reactance increases as the frequency increases, causing the current to fall. The change in reactance is proportional to the change in frequency (*see* Fig. 79). This can be simply expressed. The symbol X is given to reactance and in order to distinguish between inductive and capacitive reactance X_L and X_c are used. Reactance has the same units as resistance, i.e. the ohm. The inductive reactance $X_L = \omega L\ \Omega$, where $\omega = 2\pi f$, f is the frequency in hertz and L is the inductance in henrys.

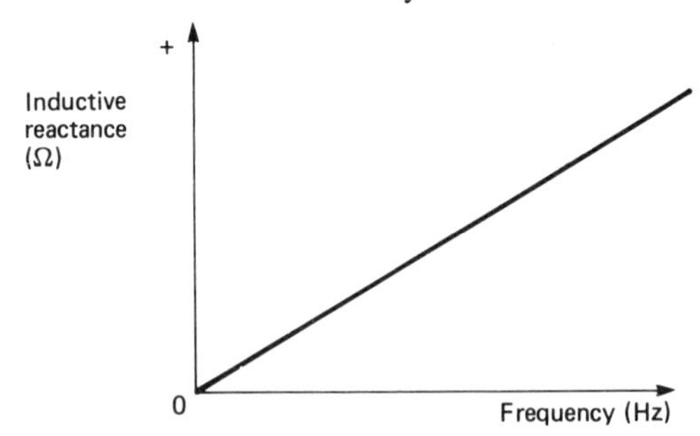

Fig. 79. *Graphical representation of the relationship between inductive reactance and frequency*

DEPENDENCE OF CAPACITIVE REACTANCE ON FREQUENCY

In the case of capacitive reactance, experiments show an opposite effect to that seen in the case of inductive reactance. It is found that as the frequency increases the current in the circuit also increases, suggesting a reduction in the opposition, i.e. in the capacitive reactance X_c. When the relationship is measured accurately it is found that the following expression governs the reactance X_c.

$$\text{Capacitive reactance } X_c = \frac{1}{\omega C}\ \Omega$$

$$= \frac{1}{2\pi f C}\ \Omega,$$

where f = frequency in hertz

and C = capacitance in farads

It is worth noting at this point that the farad is a very large unit and it is very unlikely to be met with in a practical situation. A much more practical unit is the microfarad or one millionth part of a farad, and it is necessary to convert this into the basic unit so that the expression quoted is accurate. Remember, 32 microfarads or 32 μF means 32×10^{-6} F.

The change in capacitive reactance is not proportional to frequency and is shown in the Fig. 80. Once again there is really no such thing as negative ohms, but it is convenient to express it as a negative to show its opposite nature to the inductive

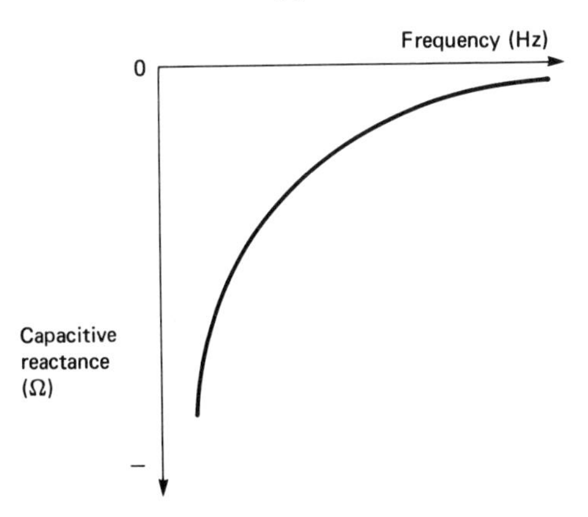

Fig. 80 *Graphical representation of the relationship between capacitive reactance and frequency*

reactance. It should be clear by now that these two components are useful in alternating current circuits, but do present opposite senses though in some ways similar characteristics.

Example 1
Calculate the reactance of a 32 mH coil when the applied voltage has a frequency of 50 Hz.

$$\begin{aligned}\text{Inductive Reactance } X_L &= 2\pi fL \\ &= 2\pi \times 50 \times 32 \times 10^{-3} \\ &= 10.05\ \Omega\end{aligned}$$

Example 2
Calculate the reactance of a capacitor of 8 μF when the applied voltage has a frequency of 50 Hz.

$$\begin{aligned}\text{Capacitive Reactance } X_c &= \frac{1}{2\pi fC} \\ &= \frac{1}{2\pi \times 50 \times 8 \times 10^{-6}} \\ &= \frac{10^6}{2\pi \times 50 \times 8} \\ &= 397.8\ \Omega.\end{aligned}$$

When the applied voltage is known the current in the circuit can be deduced by using the reactances calculated and applying the well known Ohm's law relationship.

Example 3
Calculate the current flowing in a 32 mH coil when the applied voltage is 20 V and has a frequency of 50 Hz.

From Example 1 we know that the reactance X_L is 10.05 Ω. By applying Ohm's law we can see that $I_L = V_L/X_L$ and since all of the supply voltage will appear across the coil:

$$I_L = \frac{20}{10.05} = 1.99\ \text{A}.$$

SERIES COMBINATIONS OF COMPONENTS IN a.c. CIRCUITS

The Inductance and Resistance in Series

The previous sections have dealt with "pure inductance" and "pure capacitance". Both of these are virtually impossible to achieve in practice and are simply a means of understanding the behaviour of the practical circuits involved in engineering. We can now combine the factors known about resistance in the a.c. circuit and inductance in the a.c. circuit and study how the circuit will react when both are present. A coil may be considered to be a pure inductance in series with its own physical resistance. A circuit can be set up in this way (*see* Fig. 81) and the current and voltages around the circuit can be calculated.

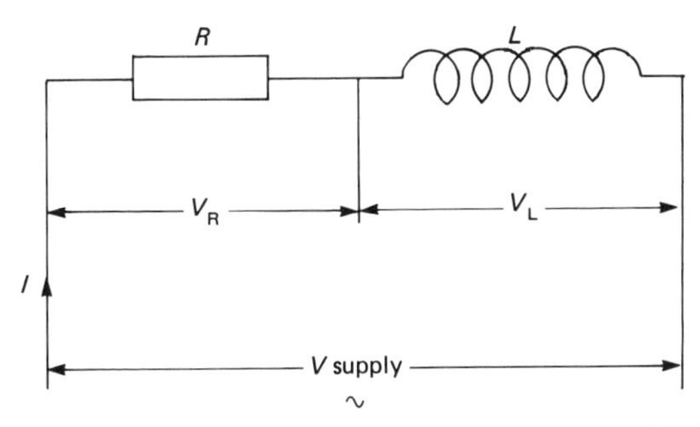

Fig. 81. *Equivalent circuit for a coil having inductance and resistance*

At first sight this may be thought to be identical to the d.c. case. The total opposition in ohms should first be calculated and used to find the current flowing in the circuit as a whole. This statement is correct but two additional factors must be kept in mind. The total opposition is a combination of the resistance, which remains constant regardless of the frequency of the supply, and a reactance which will vary as the frequency varies. Secondly, even when the two values in ohms are known, they each have a different phase relationship relative to the current which will flow around the total circuit, and direct addition will not solve the problem. These phase relationships must be taken into account before a calculation can be made. Let us look first at what we would expect to be the phasor diagram for this combined inductance and resistance circuit, remembering the principles outlines earlier.

The current in the series circuit is the common factor and this can therefore be used as the reference phasor throughout. Referring to Fig. 81 we can see that there is a voltage across the resistor which can be labelled V_R. This voltage is in phase with the current and is shown as such. The voltage V_L across the inductor however, always leads the current by 90°. It is thus shown advanced anti-clockwise by 90°. The total applied voltage

V must therefore be the sum of these two voltages but, as can be seen from Fig. 82, it is a *phasor sum* and not a simple *arithmetic sum*. The phasor sum V_S, will not be in phase with the current, but will have a phase angle φ. This phase angle will vary depending upon the size of the individual circuit components and is known as the overall circuit phase angle, i.e. that between the

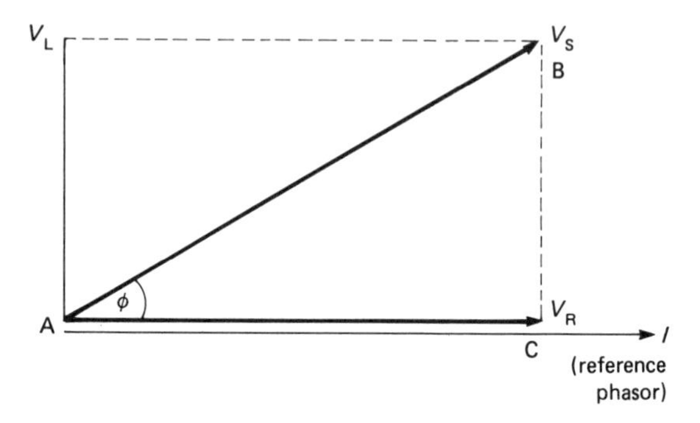

Fig. 82. *Phasor diagram for the circuit in Fig. 81*

circuit current and the supply voltage. In this case it is known as a leading or positive phase angle, which fits with the descriptions already given.

The Capacitance and Resistance in Series

A similar argument for the capacitive circuit will produce the phasor diagram shown in Fig. 83. The phase angle φ is not necessary the same as in Fig. 82 and can have any angle depending on the size of the resistive and capacitive components.

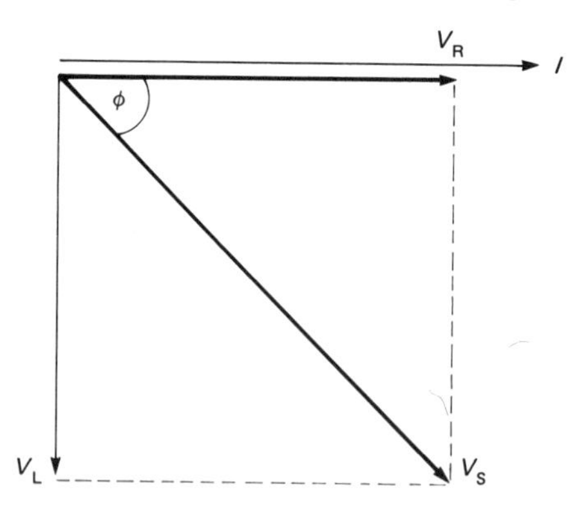

Fig. 83. *Phasor diagram for a series resistive-capacitive circuit*

In this case the phase angle is negative because the original voltage across the capacitor C lagged behind the current by 90°. Once again the total supply voltage is a phasor sum and not an arithmetic sum.

Voltage Triangle

Looking again at the phasor diagram in Fig. 82, the triangle ABC can be extracted from it and put to use in determining the various voltages and phase relationships in a circuit. The lengths of the three sides represent the in phase voltage (resistive voltage); the leading voltage (the reactive voltage) and the phasor sum voltage (which is that provided by the supply). This voltage triangle can often be very useful when solving circuit problems (*see* Fig. 84).

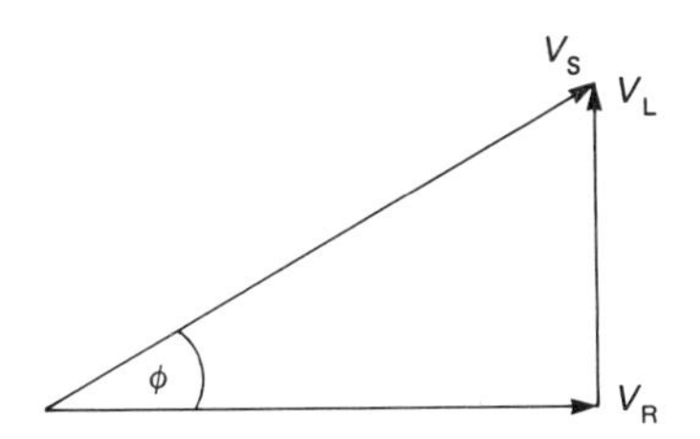

Fig. 84. *Voltage triangle for Fig. 82*

The Impedance Triangle

An extension to this voltage triangle can be made, which may also be applied to help provide circuit solutions. The current is common to all the components in the series circuit, i.e. I has the same value. Let us now perform a simple process and divide V_R, V_L and V_S by I in turn. The results are V_R/I (which, as we know from earlier work, is the resistance R; the opposition of the resistance to current flow), V_L/I (which is the reactance X_L, the opposition of the inductance to current flow) and V_S/I.

V_S, as we have noted is the phasor sum of the total voltage across the inductance and resistance. It would seem reasonable then that V_S/I is the phasor sum of the total oppositions of the inductance and resistance combined. This is accurate, but since it is neither resistance, nor is it reactance, a third term is needed to describe this overall circuit characteristic. It is known as the *impedance* (Z) and the voltage triangle can be redrawn as the *impedance triangle* (*see* Fig. 85).

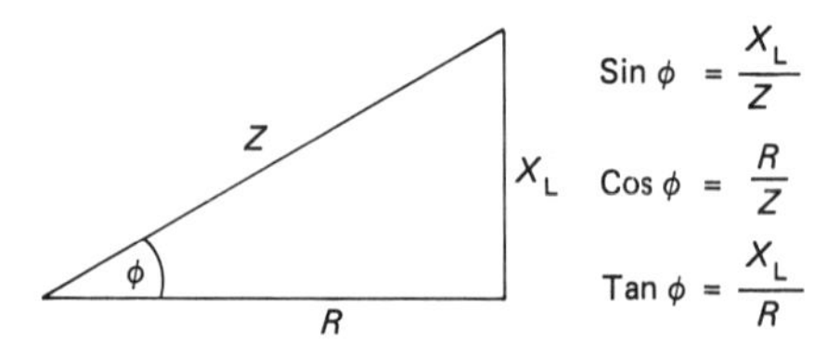

Fig. 85. *Impedance triangle for a resistive-inductive circuit*

It can be seen that provided sufficient information is known about the sides or angles, all the components of the triangle can be deduced. This is often a convenient way of calculating circuit currents and voltages.

The calculation of impedance (Z)

The application of Pythagoras' theorem to the impedance triangle gives:

$$Z^2 = R^2 + X_L^2$$

and

$$Z = \sqrt{R^2 + X_L^2}$$

Where R is the resistance in ohms and X_L is the inductive reactance in ohms.

A similar expression can be deduced when the reactance is capacitive and there is a negative phase angle.

Thus,

$$Z = \sqrt{R^2 + X_C^2}\ \Omega.$$

Example 4

Calculate the current taken from the supply when a coil of resistance 40 Ω and inductive reactance 30 Ω is connected across a 200 V a.c. supply at a frequency of 50 Hz.

$$Z = \sqrt{R^2 + X_L^2}$$

$$= \sqrt{40^2 + 30^2} = 50$$

$$I = \frac{V_S}{Z}$$

$$= \frac{200}{50} = 4\text{A}.$$

Calculation of the Phase Angle φ

Using the impedance triangle the phase angle can be calculated using any of the trigonometrical ratios, depending upon which of the triangle sides are known, i.e.

$$\tan \varphi = \frac{X_L}{R} \quad \text{or} \quad \frac{X_C}{R};$$

$$\sin \varphi = \frac{X_L}{Z} \quad \text{or} \quad \frac{X_C}{Z};$$

$$\cos \varphi = \frac{R}{Z}.$$

The ratio R/Z for $\cos \varphi$ is often the best choice since it has a particular use when calculating the power dissipated in the circuit. Cosine φ is often known as the power factor and will be used in later work.

Example 5
Calculate the phase angle in a circuit which comprises a resistor of 12 Ω in series with a capacitive reactance of 5 Ω.

The phase angle is φ, and from the impedance triangle:

$$\tan \varphi = \frac{X_C}{R} = \frac{5}{12}.$$

$$\therefore \quad \varphi = 22.6°$$

THE POWER DISSIPATION IN L–R AND L–C CIRCUITS

We have already examined the power which is dissipated in the pure inductance and capacitance and have discovered that in the pure components no power is dissipated. However, power is dissipated in the resistive components and this will depend upon the current flowing and the resistance in ohms. At all times it is the in phase component which dissipates the power and the calculation is similar to that for a d.c. circuit.

Hence, $\quad \text{power} = I^2R \text{ W}$

The Combined Effect of L, C and R Components in Series in an a.c. Circuit

Let us now consider the solution when all three components are in series in the circuit. All the principles previously explained still apply, and the overall solution depends upon the values of the components and hence their reactances. The most logical approach is via the phasor diagram shown in Fig. 86.

The usefulness of the phasor diagram now starts to become more obvious. We have already explained that the inductive and capacitive reactances are of opposite sense. This means that their own phasor sum can be calculated independently. The resultant reactive voltage is given by the difference between V_C and V_L, and appears on the phasor diagram. This becomes the total effective reactive component and it can be added to the

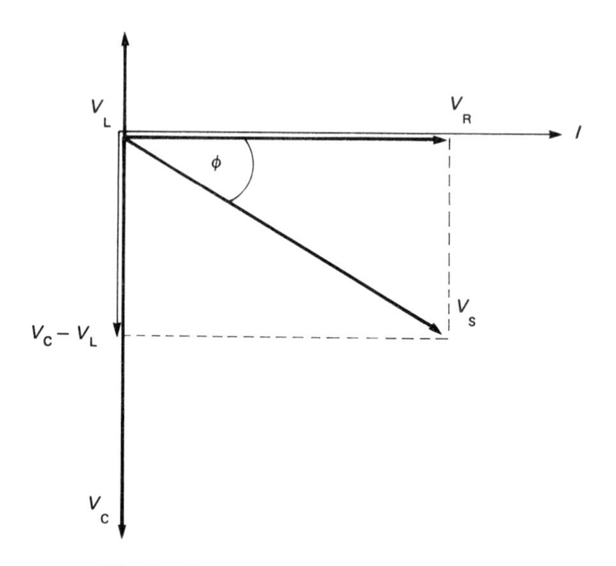

Fig. 86. *Phasor diagram for an LCR circuit*

component V_R as before to produce V_S. The rest of the circuit solution then follows from this point and can be completed as previously. A full solution for a typical circuit is given below as an example.

Example 6

A circuit consists of a 10 Ω resistor, a 0.1 H inductor and a 50 μF capacitor connected in series across a 100 V, 50 Hz supply. Calculate the circuit current.

$$\begin{aligned} X_L &= 2\pi fL \\ &= 2\pi \times 50 \times 0.1 \\ &= 31.42\ \Omega. \end{aligned}$$

$$X_C = \frac{1}{2\pi fC}$$

$$= \frac{10^6}{2\pi \times 50 \times 50}$$

$$= 63.7\ \Omega.$$

The combined impedance

$$Z = \sqrt{R^2 + (X_L - X_C)^2}$$

$$= \sqrt{10^2 + (63.7 - 31.42)^2}$$

$$= \sqrt{100 + 1041}$$

$$= \sqrt{1141}$$

$$= 33.8\ \Omega.$$

The circuit current $I = \frac{V_S}{Z}$

$$= \frac{100}{33.8}$$

$$= 2.96\ \text{A}.$$

The phasor diagram for this circuit will be as Fig. 86.

THE SERIES RESONANT CONDITION

In the previous example, the inductive reactance was greater than that of the capacitive reactance. The resultant voltage phasor was hence drawn vertically upwards, i.e. V_L was larger than V_C. However, in the original phasor diagram (*see* Fig. 85) the value V_C exceeded that of the inductive voltage V_L. The actual size depends on the circuit component values and the frequency of the supply since, as we know, frequency will affect the result.

Assuming that the circuit component values are constant, as the frequency increases the phasor V_L will increase (as X_L increases correspondingly), V_C will reduce as X_C reduces. At one particular frequency, the value of X_L and X_C will be identical and the resultant reactive component will be zero. This particular condition is known as the resonant condition and, because the circuit is a series connection of components, it is called *series resonance*.

The Phasor Diagram for the Series Resonant Condition

When $X_L = X_C$ the circuit will display certain properties which are special to the resonant condition. The circuit behaves as a pure resistance, and the phasor diagram (Fig. 87) can be directly compared with that for a simple resistance in series Fig. 71. The total circuit voltage and current will be in phase, in spite of the reactive components which are connected in the circuit.

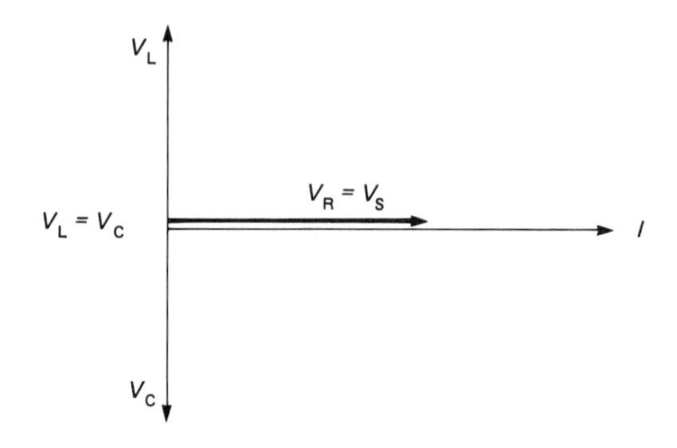

Fig. 87. *Phasor diagram for a series resonant circuit*

The phase angle is zero. The current flowing is a maximum for a given voltage. At resonance the circuit impedance will be a minimum. This can be appreciated by looking again at the expression for calculating impedance:

$$Z = \sqrt{R^2 + (X_L - X_C)^2}.$$

Thus when $(X_L - X_C) = 0$

$$Z = \sqrt{R^2}$$

$$= R.$$

The Voltages across the Reactive Components in the Series Resonant Condition

We shall now apply simple Ohm's law principles to a resonant circuit. Assume that the two reactive components chosen each produce reactance values of 50 Ω at resonance, i.e. $X_L = X_C = 50\ \Omega$, and that the d.c. resistance in the circuit totals 10 Ω. Assume a voltage of 20 V is applied at the resonant frequency of the circuit.

Let us calculate the distribution of current and voltage in the circuit as a whole.

Calculating the impedance Z:

$$Z = \sqrt{R^2 + (X_L - X_C)^2}$$
$$= \sqrt{10^2 + (50 - 50)^2}$$
$$= \sqrt{10^2}$$
$$= 10\Omega.$$

Calculating the current I_T:

$$I_T = \frac{V_T}{Z_T}$$
$$= \frac{20}{10}$$
$$= 2\text{A}.$$

Calculating the voltage across the resistance V_R:

$$V_R = I_T \times R_T$$
$$= 2 \times 10$$
$$= 20\text{ V}.$$

Calculating the voltage across the inductance V_L:

$$V_L = I_T \times X_L$$
$$= 2 \times 50$$
$$= 100\text{ V}.$$

Calculating the voltage across the capacitance V_C:

$$V_C = I_T \times X_C$$
$$= 2 \times 50$$
$$= 100\text{ V}.$$

We now have what appears to be a very unusual situation, that the voltages appearing across the two reactive components are five times as large as the applied voltage. We have normally concluded that the total voltages around a circuit will add up to the applied voltage. The rule still applies and a little examination will make it clear.

All of the examples which we have used in the a.c. section, have shown that the total voltage applied is the sum of the voltages across the components, but it is the *phasor sum* not the simple sum. This applies in the case of the resonant circuit. The 100 V in the two reactances are of opposite senses and cancel.

It must be remembered then, that very much larger voltages than the circuit applied voltages, can appear across the reactive components in the resonant condition. This fact is very important, and indeed is very often used to advantage in electronic and radio circuit engineering.

CALCULATION OF POWER IN a.c. CIRCUITS FOR SINUSOIDAL WAVE FORMS

The phase angle is once again important when we are attempting to calculate the power in the circuit. We have already looked at the power dissipation in each of the components individually and have noted that power is only dissipated in the resistive component. All of the circuit applied voltage does not appear across the resistance, but splits into two components as shown in Figs 82 and 83.

We have to calculate the power using the V_R component and the circuit current I. Power dissipated in a resistance may be calculated from:

Power $\quad P = V_R \times I$ (as in the d.c. case).

But, $\quad V_R = V_T \cos\varphi,$

and therefore $\quad P = V_T I \cos\varphi$ W.

The power can therefore be calculated using the circuit applied voltage without having to calculate the voltage across the resistance. This expression is very well known and must be remembered. The term cos φ becomes a multiplication factor, which varies depending upon the ratio of reactance to resistance. It has become known, appropriately, as the *power factor*.

When the total volts × total amperes calculation is made in the d.c. case this, apparently, is the total power being dissipated in the circuit. We have now seen that to obtain the true value we have to multiply by a factor, i.e. the power factor. The $V_T \times I_T$ expression has become known as the *apparent power* for obvious reasons, and is measured in volt-amperes. Furthermore we can see that:

$$\text{true power} = \text{apparent power} \times \text{power factor}.$$

This is a very important statement when power is considered in the a.c. circuit. In particular the power factor can be defined as:

$$\text{power factor } \cos\varphi = \frac{\text{true power}}{\text{apparent power}}.$$

The Power Triangle

Two very useful triangles have been examined in Figs. 84 and 85, namely a voltage triangle and an impedance triangle. The power triangle is a third useful alternative. The previous section explained the power dissipated in the resistive component. This can be called "effective" or "true" power and will always be shown to be in phase with the current circulating in the circuit (*see* Fig. 88). The voltage across the reactance multiplied by the

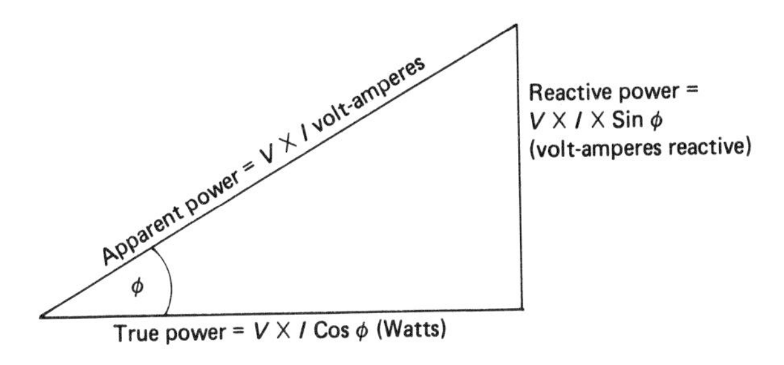

Fig. 88. *Power triangle for an inductive circuit*

circuit current would seem to cause additional power dissipation, but we have already seen that this does not happen practically. The reactive power, termed volt-amperes reactive (VA_r), is often referred to by the power engineer as VARS.

Example 7

Calculate the current flowing and the power dissipated when a 30 Ω resistance is connected in series with a 40 Ω inductive reactance across a 200 V, 50 Hz supply.

Impedance $$Z = \sqrt{R^2 + X_L^2}$$

$$= \sqrt{30^2 + 40^2} = \sqrt{2{,}500}$$

$$= 50\ \Omega.$$

Current flowing $$I = \frac{V_S}{Z} \qquad = \frac{200}{50}$$

$$= 4\ \text{A}.$$

Power factor $$\cos\varphi = \frac{R}{Z} \qquad = \frac{30}{50}$$

$$= 0.6$$

Power dissipated $P = VI\cos\varphi$

$= 200 \times 40.6 = 480$ W.

SELF-ASSESSMENT QUESTIONS

1. When the frequency of the a.c. mains voltage supply is increased, the value of resistance of a resistor in the circuit:
   (*a*) increases;
   (*b*) drops to zero;
   (*c*) remains constant;
   (*d*) decreases.
2. The opposition to the flow of current in a pure inductor is called:
   (*a*) resistance;
   (*b*) inductance;
   (*c*) inductive reactance;
   (*d*) impedance.
3. The opposition to the flow of current in a pure capacitor is called:
   (*a*) capacitance;
   (*b*) impedance;
   (*c*) resistance;
   (*d*) capacitive reactance.
4. The opposition to the flow of current in a pure inductor:
   (*a*) remains constant as the frequency increases;
   (*b*) reduces as the frequency increases;
   (*c*) increases as the frequency increases;
   (*d*) always depends on the amount of current flowing through it.
5. The opposition to the flow of current in a pure capacitor:
   (*a*) reduces as the frequency increases;
   (*b*) increases as the frequency increases;
   (*c*) remains constant as the frequency increases;
   (*d*) is identical to that of the a.c. opposition.
6. Describe briefly what is meant by the terms "out of phase" and "in phase".
7. Draw the phasor diagram relating the applied a.c. voltage and current flowing through a pure inductor.
8. Draw the phasor diagram relating the applied a.c. voltage and the current flowing through a pure capacitor.
9. When an inductor is connected in series with a resistor and an alternating voltage is applied, the total opposition to the flow of current is termed:

(*a*) resistance;
(*b*) inductive resistance;
(*c*) capacitive reactance;
(*d*) impedance.

10. The value of impedance (Z) in the above question can be calculated using the following expression:
(*a*) $Z = X_L^2 + R\ \Omega$;
(*b*) $Z = \sqrt{R^2 + X_L^2}\ \Omega$;
(*c*) $Z = \sqrt{R + X_L}\ \Omega$;
(*d*) $Z = R + X_L\ \Omega$.

11. A capacitor of capacitance 50 μF is connected across a 110 V, 50 Hz supply. Calculate the reactance and the current flowing in it.

12. A coil having a resistance of 10 Ω and an inductance of 0.05 H is connected across a 200 V, 50 Hz supply. Calculate the current and the phase difference between the current and the applied voltage.

13. A resistance of 20 Ω and a capacitance of 100 μF are connected in series across a 200 V, 50 Hz supply. Calculate:
(*a*) the current;
(*b*) the phase angle for the circuit; and
(*c*) the power dissipated.

14. Describe the basic condition known as series resonance and list the characteristics which a circuit displays when resonant.

CHAPTER SEVEN

Measurement and Measuring Devices

CHAPTER OBJECTIVES

After studying this chapter you should be able to:

* describe and explain the operation and uses, advantages and disadvantages of common measuring instruments, including moving-coil and moving-iron galvanometers, voltmeters, ohm meters and cathode ray oscilloscopes;
* describe the use of bridge and potentiometer circuits for different purposes and appreciate the accuracy of instruments of this type.

Throughout the whole of technology, it is necessary not only to know the principles of an operation, but to be able to measure the precise quantities involved. Electrical and electronic engineering is no exception, and many instruments have been developed to assist us in the measurement of current, voltage, power and resistance. It is the consideration of these instruments and the techniques of accurate measurement that concerns us in this section.

THE GENERAL REQUIREMENTS FOR INSTRUMENTS

Before studying in detail any particular type of instrument, let us first decide what it is we require to help us do the job. Firstly, we must have an indicating device. Something which responds in some way to the quantity which is to be measured, and displays it so that the operator can observe the response using his basic senses. The logical conclusion is to display it visually, and to provide a calibrated scale so that the magnitude of the response can be judged accurately.

Secondly, we must provide some form of control against which the quantity to be measured can be judged. Otherwise we will get an indication that a current is flowing in the circuit, or it isn't, and there will be no means of judging the size of the current.

Thirdly, the action of the instruments must be light and offer very little friction if accuracy is to be achieved. However, this introduces its own problem. When a pointer moves freely over a scale, responding to a very small stimulus, it will tend to overshoot the true reading because of its inertia. The needle will then slow down, stop and then retrace its path towards the correct reading. Now, the pointer will tend to overshoot going in the reverse direction, and so on. The result is that the indicator, because of its ease of response, will oscillate about the true reading. This is not satisfactory in a practical instruments and methods of damping the movement without reducing the accuracy are employed. We therefore have three basic requirements.

An instrument should:

(*a*) Produce a deflecting torque proportional to the quantity being measured;

(*b*) Produce a controlling torque to define the exact quantity; and

(*c*) Produce a means of damping to produce an efficient instrument.

We can now study how these three properties are provided in two common electrical measuring instruments, the types known as moving coil and moving iron galvanometers.

The Moving-coil Galvanometer

This type of instrument relies absolutely on the motor principle previously described. Chapter 2 explained that whenever a current is passed through a conductor it sets up its own magnetic field. When this field lies within the boundaries of a second magnetic field, a reaction occurs between them and a force is generated. If one of the sources of the two magnetic fields is free to move then it will do so.

Imagine that we can produce a coil wound on a light former, and pass the current to be measured through it. This will produce a magnetic field. Assume also that we can produce a permanent and constant static field, and introduce the coil into it. The force produced will be directly proportional to the magnitude of the current in the coil and can be used to move an indicator.

Figure 89 shows how this is achieved in practice. A is a permanent magnet, usually of some form of nickel-iron alloy. B are soft iron pole pieces and provide a shaped radial magnetic field between them. The coil C, which is wound on a light aluminium former D, is mounted between the pole pieces. It is pivoted and is free to rotate under the force produced by the interaction of the magnetic fields. The indicating pointer E is fixed to the coil and will rotate with it. However, when a current

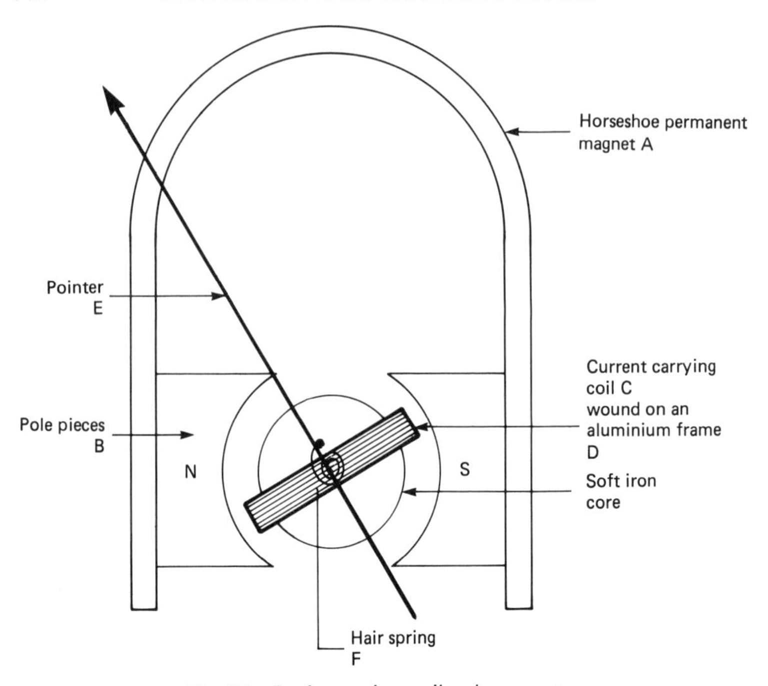

Fig. 89. *Basic moving-coil galvanometer*

is passed, provided the force is large enough to overcome the friction of the bearings, the pointer will immediately deflect to give a full scale reading and steps are taken to prevent this happening. The spring F is fixed to the coil and provides a torque which is opposite to that of the deflecting torque. As the pointer deflects, it is opposed by the controlling torque, and it will come to rest at a position where the torques are equal. In this way, the pointer will indicate different values, depending upon the strength of the deflecting torque. We already know that the deflecting torque is proportional to the current being measured and the scale can be calibrated to give the appropriate readings.

The meter will now work. All other features are to make it a better instrument, the most important of which is probably the damping already mentioned. In this case it is provided by the aluminium former *D* on which the coil is wound. Referring back to Chapter 4, we can see that whenever an e.m.f. is produced, the direction of this e.m.f. is such that it tends to oppose the effect producing it. This principle is known as Lenz's law, and applies in the meter under scrutiny.

The aluminium former is caused to rotate under the influence of the force on the coil. The former itself is effectively one low-resistance turn of conductor, and because it moves in a magnetic field, it will cut the lines of force and have an e.m.f. induced in it. Because it is a short circuit turn a current will flow in it. This current is known as an *eddy current* and will in turn produce its own magnetic field. The direction of this field, by Lenz's law, will oppose the movement producing it, i.e. the rotation of the coil. The net effect will be to steady the pointer movement, which we call *damping*. It is an efficient method and is widely used in moving coil instruments. When the pointer comes to rest, no eddy currents will be generated, and the accuracy of the meter will be unimpaired. Because of the principle involved this form of damping is called *eddy current damping*.

Moving-iron Instruments

The basic law of magnetism is that like poles repel, unlike poles attract. Imagine that two small steel rods are placed side by side inside a coil. A current is made to flow throught the coil and the rods will be magnetised with the same polarities at their ends (*see* Fig. 90). The natural reaction will be that the rods are forced apart. The size of the force pushing them apart will depend upon the size of the current in the coil and once more we have a principle which will provide us with a current measuring instrument.

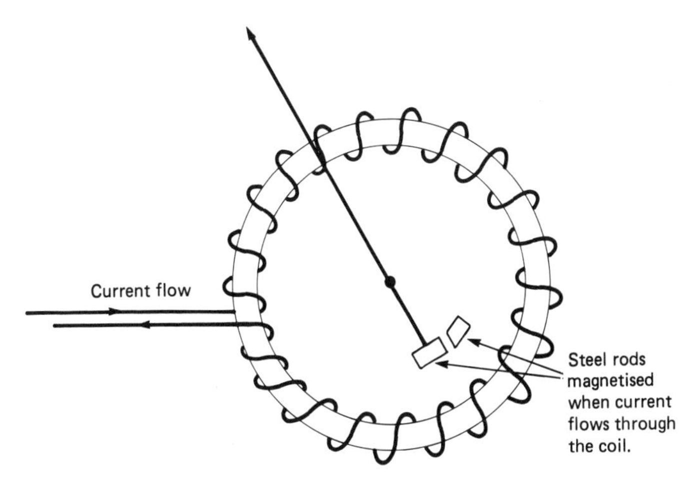

Fig. 90. *Basic principle of the repulsion type moving-iron meter*

The Repulsion type Moving-iron Meter

Figure 91 shows the basic make up of a repulsion type moving-iron meter. The current to be measured passes through the coil A. The induced magnetism in the two irons B and C is of like polarity and, as explained in the previous paragraph, gives rise to a repulsive force. The pointer D is connected to the moveable iron B indicates a deflection over the scale F.

An important difference between the moving-iron and moving-coil galvanometers is in the scale. The force between magnetised

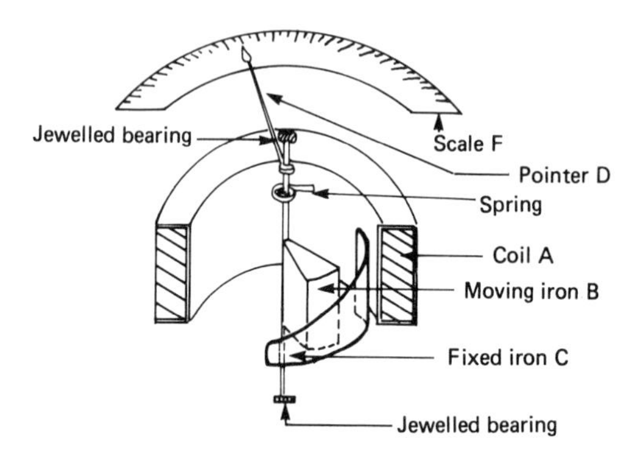

Fig. 91. *Repulsion type moving-iron meter*

bodies is inversely proportional to the square of the distance between them. Hence, the force reduces considerably as the two magnetised rods are parted. This means that the scale changes as the value of current changes. Moving-iron meters tend to have scales which are "bunched up" at the lower end (*see* Fig. 92).

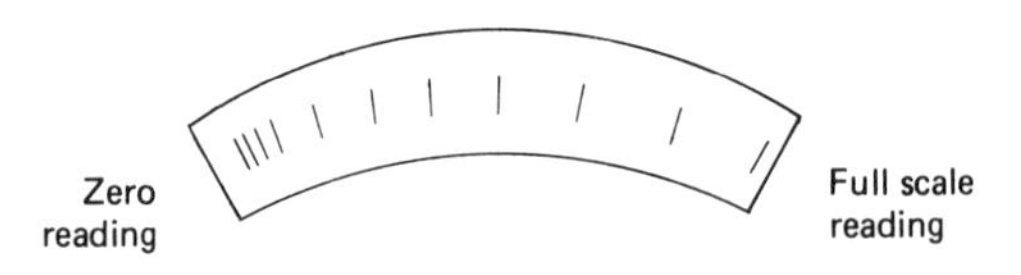

Fig. 92. *Typical scale for a moving-iron meter showing closeness of graduations for low readings*

This is one of the main disadvantages of this type of instrument and wherever possible this disadvantage is reduced by shaping the irons. An example of the modified shape of the scale is shown in Fig. 93.

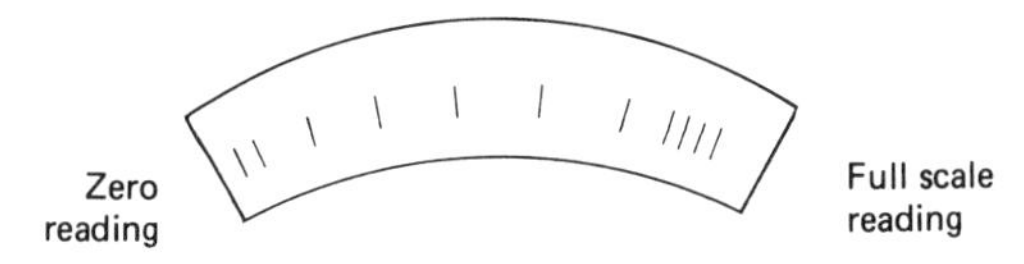

Fig. 93. *Scale for a moving-iron meter with shaped iron rods*

Control in the Moving-iron Meter

The control torque is most often provided by springs which are similar in make-up to those of the moving-coil meter. A further point is that very often two springs, made of invar, are incorporated. (Invar is used because of its electrical and mechanical properties. It has good temperature characteristics, is springy and a good electrical conductor.) This occurs in both types of meter. The main reason is to avoid meter errors due to changes in temperature. The two springs are wound in opposite directions but are balanced to provide the correct control torque. Any expansion or contraction due to changes in temperature tends to produce opposite effects in the two springs, and minimise errors which may thus be introduced.

THE CORRECT CONNECTION OF AMMETERS AND VOLTMETERS

In Chapter 1 we saw that all of the current flows through each component in a series connected circuit. It seems reasonable then, that when it is required to measure current flowing in the circuit, a series connection is made. The circuit is physically broken and the ammeter connected as in Fig. 94. Ideally then, the d.c. resistance of an ammeter is kept as low as possible so as not to upset the circuit in which the measurements are being made. However, when it is required to measure the potential

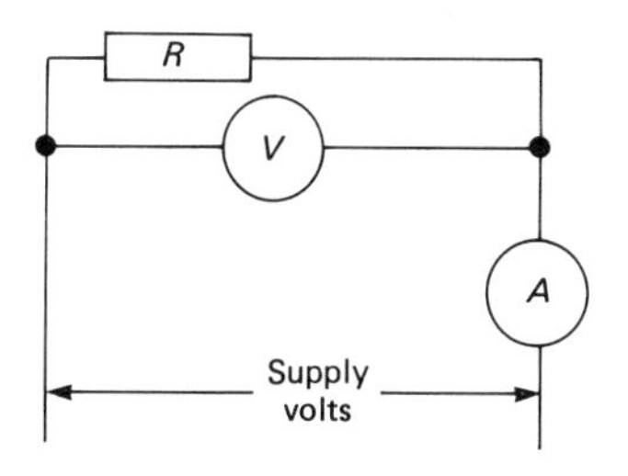

Fig. 94. *Voltmeter and ammeter connections showing ammeter error*

difference across a component or circuit, the component connections need not be disturbed. The voltmeter is connected physically across the component. Figure 95 shows the technique used to measure the potential difference in volts across *R*. The opposite ideal to the ammeter applies in this case. It is required to keep the resistance of the meter as high as possible, so that the effect of connecting a parallel resistance across a component is kept to a minimum.

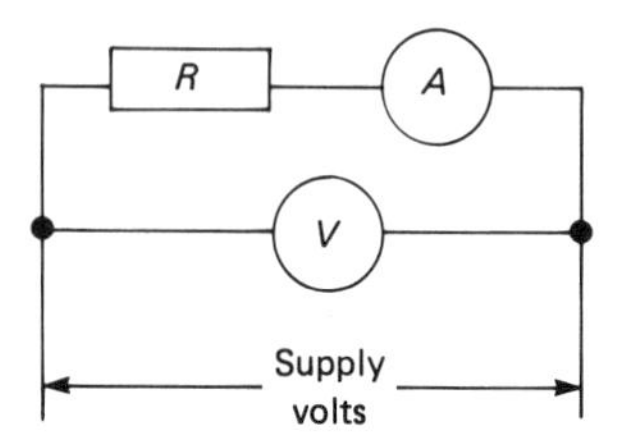

Fig. 95. *Voltmeter and ammeter connections showing voltmeter error*

MEASUREMENT ERRORS

When measurements of all types are being made, the ultimate goal is accuracy. When readings are taken in practice, several factors may reduce the overall accuracy, and we must now consider some of these.

Systematic Errors

It has been explained in a previous paragraph how ammeters and voltmeters are connected, but when both measurements are needed errors may be introduced. Examine Figs. 94 and 95. Both meters are connected correctly in their respective cases. However in Fig. 94 the ammeter will indicate the true current in the load and the small amount of current drawn through the voltmeter coil. It is this current, which although small, will produce an error.

In Fig. 95 the correct load current will be indicated on the ammeter, but the voltage across the load and across the ammeter movement will be indicated by the voltmeter. In this case this produces the error.

When it is the actual technique of measuring which introduces the error, as in these examples, the errors are known as systematic errors. Neither connection will give the correct current and voltage reading at the same time, and if accuracy is required the operator must be aware of the difficulties.

Observational Errors

A second type of human error which may be introduced, is in the observation of the indicated reading. Imagine that you drive along in front of a large clock on a public building. When the clock indicates exactly half past the hour, as you pass it from left to right the reading which you observe will apparently change from approximately 29 minutes past the hour, to 29 minutes to the hour. Also the further the large hand is away from the clock face, the more will the reading apparently change.

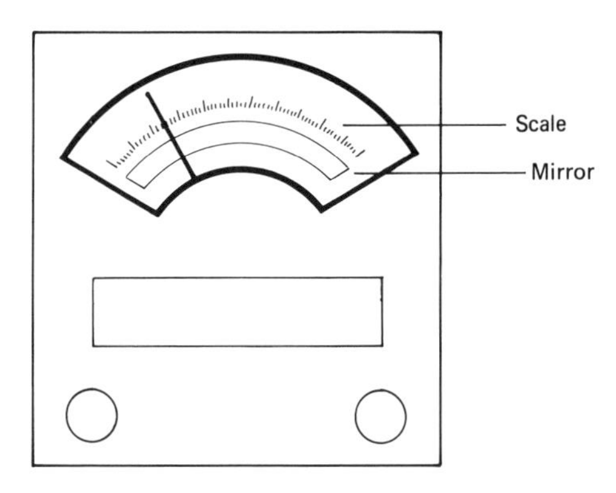

Fig. 96. *Use of a mirror to obtain accurate meter readings*

It is fairly obvious then, that the position of the person reading a meter is important, and also that the pointer and scale are finely constructed.

To help improve observational accuracy, meters often have a slim mirror behind the pointer. The observer lines up his eye so that the pointer and its image coincide. When they do, the eye will be directly above the pointer and the most accurate reading can be made (*see* Fig. 96).

Calibration Errors

Very accurate instruments are very expensive. The normal range of industrial instruments does not necessarily indicate the true reading. Typical accuracy may be quoted at ± 2% of full scale deflection. This means that for a meter with a full scale deflection of say 100 mV, any reading may well be 2 mV greater or less than the true reading. Indeed, this error may well change over the full range of the instrument and, if a high degree of accuracy is needed, the instrument must be checked across the full scale against a series of known true currents and the calibration errors mapped. Future indications can then be corrected using the error graph.

EXTENSION OF METER RANGES

The current which will cause a full scale deflection (f.s.d.) in a moving coil meter, may well be quite small (typically of the order of 1 mA). When measurements of several amperes are required, a heavier guage of wire would be needed in the coil to avoid damage and the moving coil itself would be far too large and heavy to operate. An alternative method must be found to measure these larger currents.

Current Shunts

Imagine that you have to count the number of people in a large crowd. You could ask them all to file through a gate and physically count them. Another method would be to tell them to form themselves into groups of five and, instead of all five of them filing through the gate, one from each group is to come through and the other four are to leave by another gate. You can then count the singles and, knowing that the 4:1 ratio is constant, you can convert your "reading" to the correct figure. Larger ratios can be used for larger crowds.

This is exactly the principle which we use to measure the larger currents. An "alternative gate" is provided. A resistor is connected in parallel with the meter movement and the larger proportion of the current goes through it. This leaves only an amount which the meter movement can cope with comfortably. The Fig. 97 shows the arrangement. The name given to this parallel resistor, is a *shunt*. We have already said that ammeter movements must have a low resistance. Therefore the shunt, to allow most of the current to pass through it, must have a very low resistance. Resistances of 0.001 Ω, or even lower, are quite typical.

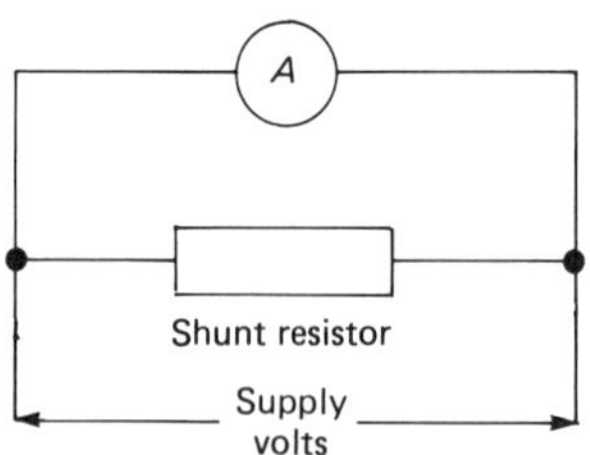

Fig. 97 *Resistor used as a shunt to extend the range of an ammeter*

Voltage Multipliers

In a series circuit, the total voltage applied produces potential differences across all the components in the circuit in proportion to their resistances. The larger the resistance, the larger the voltage drop across it. This gives us a ready method of extending the range of a voltmeter. The f.s.d. for the voltmeter may be say 10 V, but it may be required to measure voltages up to 100 V. Provided that we know the resistance of the meter movement, an extra resistor can be added, this time in series, to "protect" the meter. Figure 98 shows the idea. The extra resistor effectively

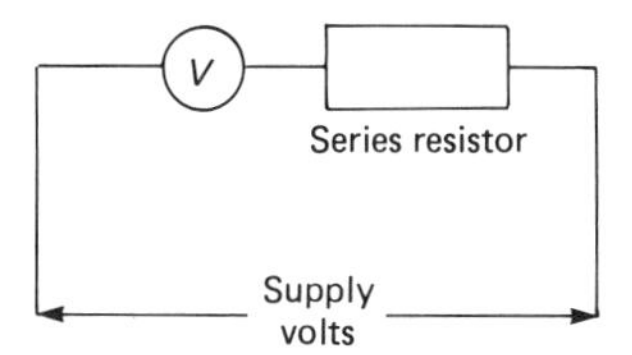

Fig. 98. *Resistor used as a multiplier to extend the range of a voltmeter*

multiplies the range of the meter and when used in this way is called a *multiplier*.

Voltmeter movements have high resistances, therefore to be effective the multiplier will need to have a very high resistance indeed. Resistances of the order of 1 MΩ or higher are typical for practical multipliers.

THE CATHODE RAY OSCILLOSCOPE AS A MEASURING DEVICE

We have recently considered the problem of errors in measurements, and have deduced that the input resistance of a voltmeter should be very high, so that it draws very little current from the supply. An ideal voltmeter would have an infinite resistance.

The oscilloscope can be used to measure a.c. or d.c. voltages without a great deal of difficulty. The oscilloscope is fed internally with a signal of set level, and this can be used to calibrate the amount of beam deflection for a given voltage. An a.c. voltage to be measured can be fed to the input terminals and "locked" to produce a visual waveform. Using the graticule provided, the height of the peak voltage and the time-scale of the waveform can be read directly from the trace. The results are then adjusted by taking the calibration of the vertical and horizontal components of the display into account. Measurements of

voltage, period and frequency can be made quite easily by using this technique.

The input circuitry of the C.R.O. offers a very high resistance and so the device fulfils the basic requirement for a good voltmeter. The connections are made in exactly the same way as they would be for a conventional voltmeter, connecting the oscilloscope in parallel with the component the voltage drop across which is to be measured.

Where it is necessary to measure d.c., as well as a.c. voltages, the C.R.O. can be coupled internally to take notice of, or ignore, the d.c. level by throwing an a.c./d.c. switch. The net result is to move the trace vertically, the amount of deflection indicating the d.c. level. Again, account must be taken of the setting of the vertical control to obtain the correct value.

THE LIMITATIONS OF MOVING-COIL AND MOVING-IRON METERS

A current passing through a moving coil in the correct direction, will produce an indication on the scale. A reversal of this current, however causes the movement to hit against the back-stop and no indication is given. This means that a rapid reversal of current (a.c. current) cannot be measured directly using a moving coil meter.

The device can be modified by rectifying the a.c. current to be measured, and feeding the resulting d.c. to the meter coil. The scale must then be adapted to give the correct readings. The section on rectified a.c. (Chapter 6) shows that the output waveform is a series of half-sine waves, and the meter deflection will be proportional to the average current flowing. This can be shown to be $0.637 \times I_{max}$, and the deflection will indicate that value. When a meter is required to indicate r.m.s. values, the scale will have to be adapted by the factor $0.707/0.637 = 1.11$. It must be noted that this will only be accurate for sine waveforms, and care must be taken where non-sinusoidal a.c. is being measured.

The opposite is true for the moving-iron meter. In this case the deflection is proportional to I^2 and since $(-I)^2 \equiv I^2$ this will measure a.c. directly. Also, since the deflection will be proportional to the average or mean of the I^2 values the meter will read the r.m.s. value. This is a further advantage.

The moving-iron meter is mainly used for low frequency a.c. applications and is not used with shunts or multipliers, in contrast to the moving coil meter. In this case the magnetising coil can be wound with much thicker wire for high current applications, without affecting the movement.

NULL MEASUREMENT TECHNIQUES

A very important technique is that of recognising a minimum measurement. When direct readings are made, they may well include contributions from sources other than the one which it is important to measure. This has given rise to techniques which compare the unknown signal with a known signal. In this way extra voltage or current, due to say, temperature effects occurs in both the wanted and unwanted measurement, and tend to cancel each other out, increasing the accuracy of the measurement made. The known signal is fed to some form of indicating device via accurately calibrated controls. The unknown signal is also fed to the indicating device but through circuitry opposite in sense, i.e. phase or polarity. Naturally, when the two are equal a null reading is obtained. The signal fed via the controls is varied until this null, or minimum, reading is obtained. When this is the case, the equivalent values can be read from the settings of the controls. Two excellent but simple examples of null reading methods are described in the following paragraphs.

THE WHEATSTONE BRIDGE

Figure 99 shows a Wheatstone bridge. The principle is extremely simple. When B and C are at the same potential no current will flow through the meter E, there is a null reading. The currents I_1 and I_2 must give equal potential differences across the resistors P and Q if B and C are to be of equipotential. Hence $I_1P = I_2Q$,

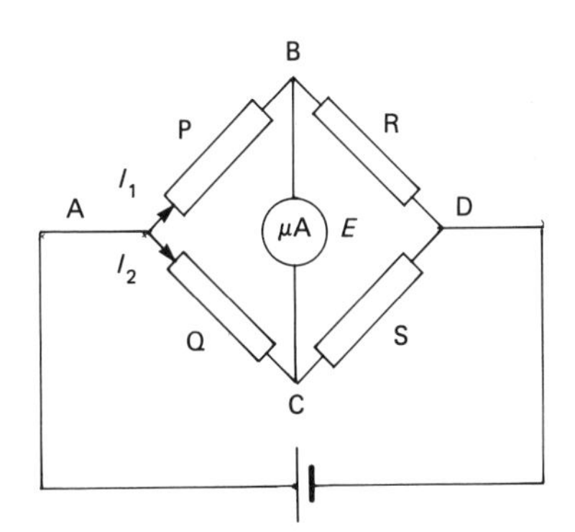

Fig. 99. *The Wheatstone bridge*

where P and Q are the resistances of the resistors P and Q in ohms. Now, the point D is common, meaning that the potential differences across R and S must also be equal and $I_1R = I_2S$, where R and S are the resistances of the resistors R and S in ohms. We can divide the first equation by the second to get equal ratios:

$$\frac{I_1P}{I_1R} = \frac{I_2Q}{I_2S}$$

$$\therefore \qquad \frac{P}{R} = \frac{Q}{S}$$

$$\text{and} \qquad P = Q \times \frac{R}{S}.$$

Hence Q is made an accurately calibrated variable resistor and R and S are made equal, the unknown resistance P is placed in the bridge circuit and the battery connected. Q is varied until a minimum (in this case a null) reading is obtained on the meter E. The resistance of Q then must be equal in value to the resistance of the unknown resistor P.

The Ratio Arms R and S

In the example given the limit values of Q must cover the values of R or a zero reading cannot be obtained. This is a limitation of the device and steps can be taken to increase the range by varying the resistances of R and S. A common method used is to make R and S variable by factors of 10. In this way the values for Q, which may range from 1–10 Ω can be extended to cover 0.1–1 Ω and 10–100 Ω simply by changing the ratio of R and S.

In a practical Wheatstone bridge R and S have become known as the ratio arms for obvious reasons. They may cover ratios of 100:1 or even 1,000:1, giving a very wide useful range of resistance measurement.

THE d.c. POTENTIOMETER

Let us revise the basic principles governing the connecting of simple cells in series. Figure 100(*a*) shows two cells connected in series aiding. These will obviously give rise to a deflection on the meter. Figure 100(*b*) shows two cells of different e.m.f. connected in opposition. In this case the meter will give a reading but its size will depend upon by how much the e.m.f. of one cell exceeds that of the other. If it so happens that the two values of

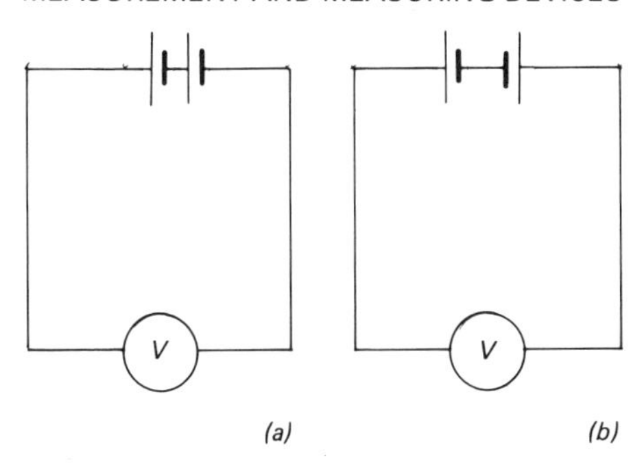

Fig. 100. *Cells connected in series:* (a) *aiding; and* (b) *opposing*

e.m.f. are equal, then the meter will give a zero reading. This simple principle can be used to produce another null measuring technique.

Figure 101 shows a simple cell of e.m.f. 2 V connected across a wire P of uniform cross section and length L. It is reasonable to assume that half of the battery e.m.f. is dropped along half the length of the wire, and that the potential drop from R to Q (which is half-way between R and T) is 1 V. The length l (in this case $l = \frac{1}{2} L$) effectively represents a potential drop of 1 V. Assume now that a cell X of e.m.f. 1 V is connected at R and is in series with a centre zero galvanometer or milliameter attached to a wander lead as shown in Fig. 102. The wander lead can be touched to the wire P at any point Y. When contact takes place the two e.m.f.s in the length of wire RY are compared. They are opposite in polarity and, depending on which is the greater, the meter deflection will be in one or other direction. The position of

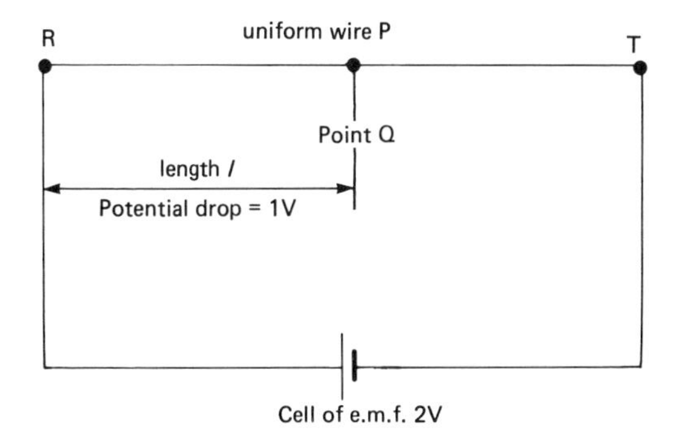

Fig. 101. *Uniform wire* P *connected to a simple cell*

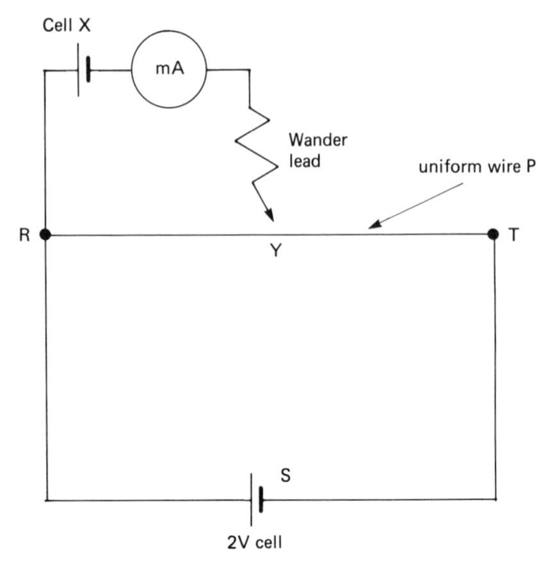

Fig. 102. *Simple method of null measurement using a potentiometer*

Y can be varied until the meter gives a null reading, indicating that the two e.m.f.s in the length of wire RY have been balanced. In this case, of course, the balance point will be at RY = $\frac{1}{2}L$.

The d.c. potentiometer makes use of this principle. Firstly, a cell with a very accurately known e.m.f. E_1 is connected at R and the wander lead moved up and down the resistance wire R until the null point is found. The length l_1, measured on a scale placed under the wire, then represents the e.m.f. of the known cell. A cell whose e.m.f. E_2 is to be measured is then substituted at R, and the new null position found by re-adjustment of the wander lead. This length can be labelled l_2. A direct comparison of l_1 and l_2 will deduce the value of the unknown e.m.f. by a simple ratio.

The accuracy of this method will depend upon the exact voltage of the original cell. However, standard cadmium cells, giving exactly 1.0186 volts are available which introduces a high degree of accuracy overall, and the method is very acceptable.

The Practical Potentiometer

A long straight length of resistance wire is very cumbersome and the practical potentiometer has a variable resistance in a much more compact form. Nevertheless, very accurate reading of e.m.f. can be made up to the value of the cell S. In addition the device can be extended to measure current by measuring accurately the voltage a cross a standard resistance of exactly 1 Ω. The device suffers from the further disadvantage in that it does not give a direct reading and a calculation has to be made each time.

Resistance Measurement using the Potentiometer

A further change can be made to enable us to measure an unknown resistance R_x. In this case the unknown resistance is connected in series with a known resistor of resistance R_k and a cell (*see* Fig. 103(*a*)).

The potential differences V_x and V_k across each of the resistors are balanced against the potentiometer wire lengths l_x and l_k, as shown in Fig. 103(*b*).

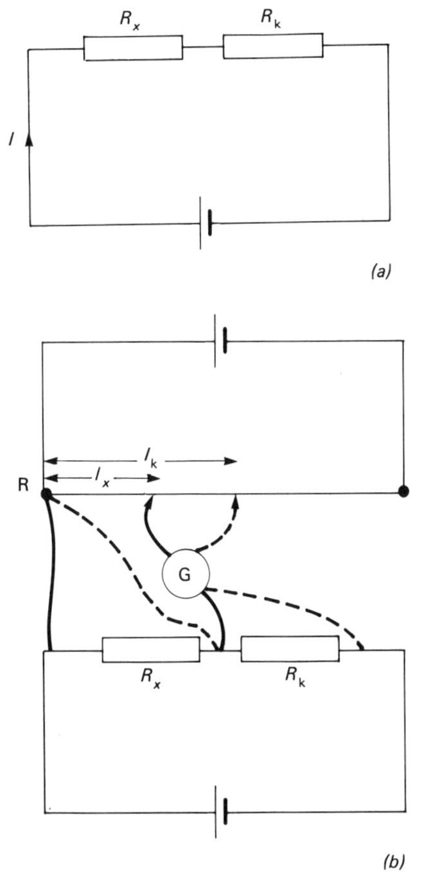

Fig. 103. *Resistance measurement by means of a potentiometer*

We know that $$\frac{V_x}{V_k} = \frac{l_x}{l_k}.$$

From Ohm's law $$V = IR$$

Therefore, $$\frac{I_x R_x}{I_k R_k} = \frac{l_x}{l_k}.$$

However, since we know that the current flowing in both resistors must be the same, we have obtained a simple ratio from which the unknown resistance can be calculated.

THE OHM METER

This is a device for measuring resistances directly. This type of facility is usually included in multimeters. It uses an identical meter movement to that of the ammeter or voltmeter, but with the zero at the opposite end of the scale. Figure 104 shows how the three scales of a multimeter may be arranged.

The circuit for the ohm meter is very simple. It consists of a source of e.m.f., which is usually a replaceable primary cell included inside the body of the multimeter. In series with it is the meter movement and a variable resistor R which is used to adjust the zero for accurate readings. The unknown resistance to be measured is connected across two terminals which are labelled A and B.

The full sequence of measurement is as follows:

1. The terminals A and B are first short circuited and the resistor R adjusted to give zero reading on the scale. This compensates for any possible slight changes in cell voltage, etc. The resistor R is usually labelled zero adjust on the face of the meter.

2. The short circuit is removed and the unknown resistor to be measured is connected directly across the terminals A and B.

3. The resistance scale is then read directly and the value of resistance noted.

Ohm's Law tells us that as the value of the unknown resistor increases, the current through the circuit will reduce, and the meter deflection will decrease. A much smaller value of unknown resistance will allow a larger current to flow and the deflection will increase. Hence, the need for a "reversed" scale can easily be understood.

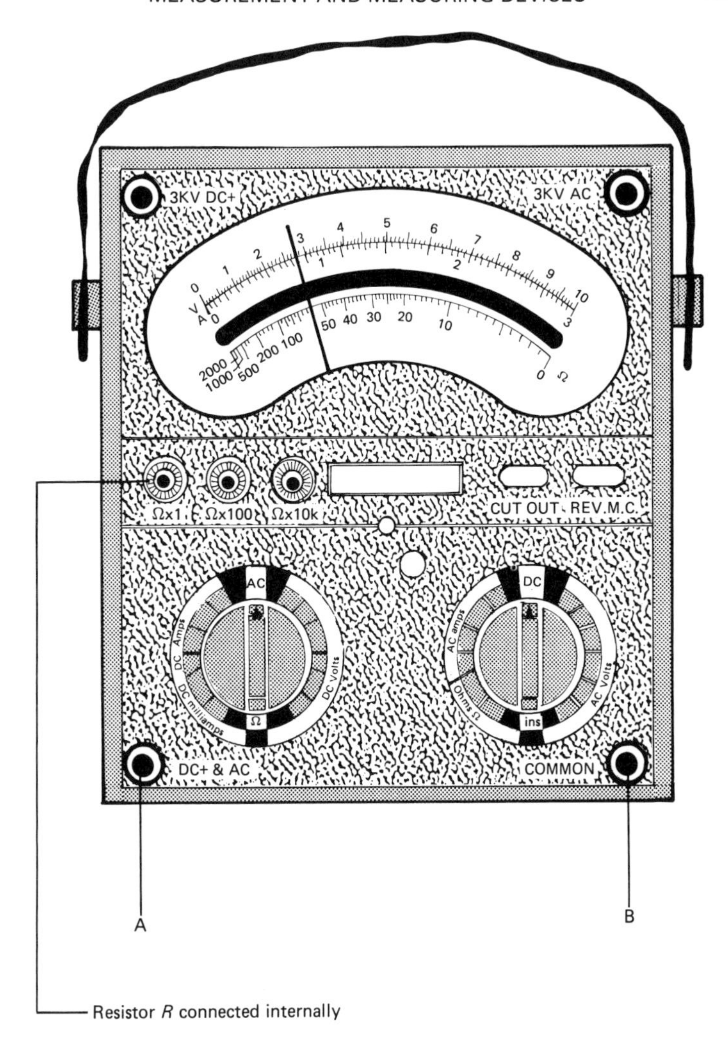

Fig. 104. *A commercial multimeter showing a typical arrangement of scales*

Extension of the Resistance Range

Looking at the high end of the resistance scale in Fig. 104, we can see that at the top end of the scale the readings become increasingly inaccurate. In order to try to counteract this the internal adjust resistance R, often consists of three separate resistors which are switched in or out of the circuit by means of a range change switch.

The three positions of the switch set up the meter to read:

(*a*) "Ω × 1" which means the scale reading is read directly;

(*b*) "Ω × 100" meaning that the scale reading is multiplied by 100 to give the correct value; and

(*c*) "Ω ÷ 100" indicating that the scale reading must be divided by 100 to give the correct value.

A suitable choice of range enables much wider range of values of unknown resistance to be accurately measured.

DIGITAL METERS

All of the meters which we have previously considered have needed the operator to use his skill and judgment in obtaining the required reading. The modern trend is towards instruments which have the measured quantity displayed in numerical form. Modern integrated circuit technology which provides this output form will produce readings to great degrees of accuracy. It is likely that, as integrated and microcircuits become cheaper, all meters will eventually have displays of this type.

SELF-ASSESSMENT QUESTIONS

1. In order to convert a milliameter into a voltmeter it is necessary to connect:

(*a*) a large resistance in parallel with the instrument;
(*b*) a small resistance in parallel with the instrument;
(*c*) a large resistance in series with the instrument;
(*d*) a small resistance in series with the instrument.

2. With the aid of a diagram explain the principle of operation of the moving-coil galvanometer. Explain how the deflecting torque is produced and how control is effected. What advantages and disadvantages does this instrument possess compared with the moving-iron instrument.

3. A moving-coil galvanometer has a resistance of 5 Ω and takes a current of 10 mA to give full-scale deflection. Draw a diagram to show how a resistor could be connected to enable the instrument to give a full-scale deflection for a current of 1 A. Calculate the value of resistance required.

4. A current flows through a resistance of 10 Ω and the p.d. across the resistor is measured on a potentiometer a balance being found at a distance of 25.45 cm. If a standard cell of e.m.f.

1.0186 V balances at a distance of 50.9 cm, calculate the value of current in the 10 Ω resistor.

5. Explain the operation of the Wheatstone bridge and develop the balance equations. The diagram shows a bridge circuit. Calculate the value of the unknown resistor needed to produce a balance on the galvanometer.

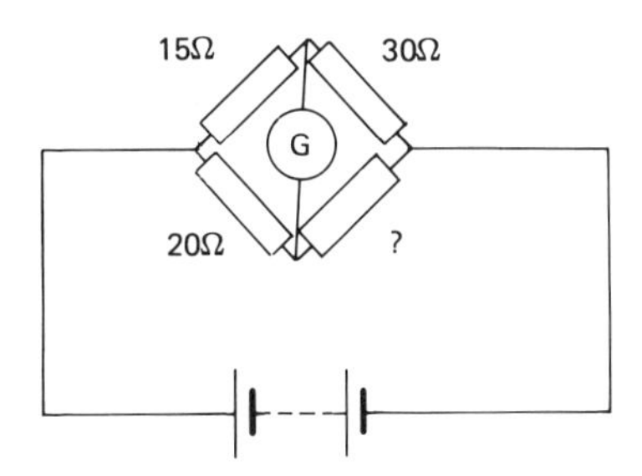

6. A meter having a resistance of 15 Ω gives full-scale deflection with 5 mA flowing through it. Calculate the values of the additional resistances required to convert the instrument to:

(*a*) an ammeter measuring 50 mA at full-scale deflection;
(*b*) a voltmeter measuring 50 V at full-scale deflection;
Show the method of connection in each case.

7. A moving-coil galvanometer has a resistance of 4 Ω and a current of 15 mA produces full-scale deflection on it. It is required to convert the instrument into a voltmeter having three ranges with full-scale deflections of 50 V, 150 V and 300 V. Calculate the value of the additional resistance required in each case and draw a diagram to show how the resistance would be connected, indicating the tapping points on the resistor to accommodate the various scales.

8. Discuss briefly the criteria which make moving-iron and moving-coil instruments suitable or unsuitable for different conditions of measurement in a.c. and d.c. circuits.

State which type of instrument would be the most suitable in the following cases, giving reasons for your choice:

(*a*) measurement of d.c. voltage covering the full range indicated including very small values;
(*b*) measurement of alternating currents of varying waveforms;
(*c*) measurement of alternating voltages at different frequencies from 20 Hz to 250 Hz.

CHAPTER EIGHT

Semiconductor Diodes and Transistors

CHAPTER OBJECTIVES

After studying this chapter you should be able to:

* understand the electrical and physical differences between materials which are classified as good conductors, insulators and semiconductors;
* appreciate the effects of temperature changes on these materials;
* understand how the doping of pure germanium and silicon significantly changes their electrical properties;
* understand the basic operational principle of a pn junction;
* understand the operation and characteristics of a pnp or npn transistor.

When a current flows in a material a movement of charges takes place. Electrons leave their parent atoms, and move about inside the atomic structure. Engineers contribute a great deal to everyday life by controlling this movement and making use of it to perform many functions.

ATOMIC STRUCTURE

Imagine a grain of salt. It looks like salt, tastes like salt and behaves chemically as salt does. Its chemical name is sodium chloride. Imagine that we can break the grain of sodium chloride into smaller and smaller pieces. Eventually we shall finish up with single molecules of salt. Assume we now break one of these molecules open. Inside we find two atoms, one of chlorine and one of sodium. Both these substances are elements and behave very differently if they are not compounded together. Chlorine is a gas and sodium is a metal. Imagine that we can take the individual atoms of these elements and break them open to examine what is inside them. We discover some very interesting facts. Although very different as complete atoms, inside the atom are identical pieces. There is a central body called the nucleus, which contains positively charged particles known as protons and neutrally charged particles known as neutrons. Around this central body are much smaller particles circling with a motion which is similar to the motion of the Earth and the other planets circling our sun. These particles are called electrons and have a negative charge. What is the difference, then, between the chlorine and the sodium atoms, if the particles inside each are identical? The difference lies in the number of each type of particle. Indeed, it is true to say that all of the elements are made

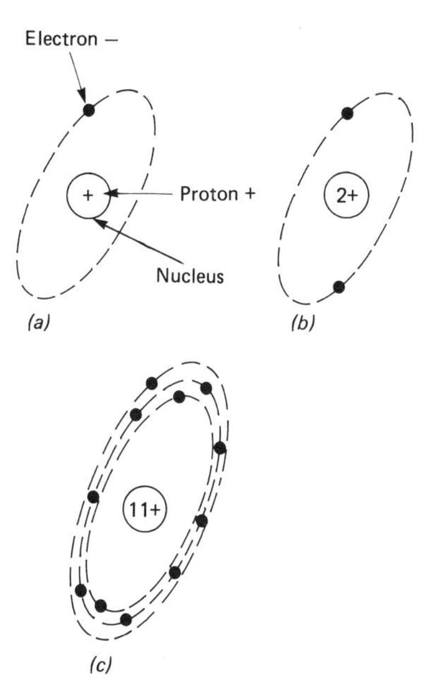

Fig. 105. *Basic atomic structures of:* (a) *hydrogen (with one electron and one proton);* (b) *helium (with two electrons, two protons and two neutrons);* (c) *sodium (with eleven electrons, eleven protons and twelve neutrons)*

of the same basic pieces but they all behave differently because of the number of each type of particle and the way they are arranged inside the atom.

Figure 105(*a*) shows the make-up of the simplest of all the elements, hydrogen. One electron circles the central nucleus which has one proton. Figure 105(*b*) shows the second most simple atom, that of the gas helium. In this case two electrons orbit around the nucleus which consists of two protons and two neutrons. The increase in complexity continues from element to element and Fig. 105(*c*) shows an atom of sodium which was mentioned earlier. Sodium happens to be number 11 in the list of elements and therefore has 12 neutrons and 11 protons in the nucleus, and 11 electrons in orbit around it.

It is important that we now note two important facts which will be of great help in the understanding of semiconductors. The first is that in a complete atom there are always as many positively charged protons as there are negatively charged electrons. This is always true, and because of this fact a complete atom is always neutrally charged. The positive and negative charges being equal in number, will cancel the effect of each other. The second point to note is that the electron orbits follow clearly defined routes. Each of these routes is called a shell, but it is found that only so many of the electrons in each atom can orbit within a particular shell. Once there are more electrons than this maximum number, a new shell is started. Figure 105 shows this very well. The first shell can take only two electrons and a second shell is started for the third electron. This will accept the next seven but no more. Thus, when we get to the eleventh electron in the sodium atom, a third shell is needed and so on. We do not really have to consider why this is so, but its effect is very important to the electrical and electronic engineer. The central nucleus exerts a holding force on these electrons, preventing them from flying away from the atom. However, when one electron exists by istelf in an outer shell, this holding force is much less strong and the electron can quite easily "escape", from the parent atom. When this happens the electron exists freely inside the material structure, and if some external force is applied they can be encouraged to move in a particular direction within the material. The external force may be an electromotive force, e.g. a simple battery. It is this movement of the negatively charged electrons which we call electric current. When this takes place inside a material, the material is said to conduct electricity, and it is termed a conductor. How well particular materials conduct electricity is what we must consider next.

Good and Bad Conductors

Having followed the explanation of atomic structure through, we must now see how the structures of different types of materials relate to how they react when an e.m.f. is applied to them. If we carry out this check on several elements such as copper, iron silver, sulphur, etc. we shall find that good conduction takes place in the case of all of the metals, but where sulphur and similar substances are concerned very poor conduction is apparent and a very low current flows. We can now examine carefully the structure of the atoms of these particular elements and do a bit of detective work. We shall find that in the case of all the good conductors, only one or two electrons exist in the outermost shell. When the atoms of the elements which show poor conducting properties are examined, their outer shells are seen to have near to the maximum number of electrons which they can tolerate. Perhaps this could be a possible explanation for their different conducting properties! Although we do not have to understand why, we have to remember a brief explanation. This effect depends upon the holding force exerted by the nucleus on the orbiting electrons. When a shell is full, or almost full, this holding force is strong. The electrons find great difficulty in breaking away from the parent atom and are therefore not available to move about within the structure and form a current. It requires a very large external force to drag them out of their orbital shell. However, it is important to note that it can happen. When a very large e.m.f. of, say, several thousand volts is applied, the electrons are forced to move away from their parent atom and are made to travel through the material as a current.

Exactly the opposite is found in the case of the metals. Their outer shells have few electrons, which are very weakly held and readily break away in large numbers. In the case of copper or silver, the electrons often move away from the parent atom without any externally applied force and are readily available in the structure to form a current.

At this point it is very important to emphasise that the action of forming an electric current is the same in each case. It is simply the ease with which one is achieved which varies. We tend to put the different materials into different boxes and call them conductors or insulators i.e. materials which will conduct and those which prevent conduction. This unfortunately suggests two different actions but, as we have already seen, this is not so, and it is best always to consider all materials to be capable of conducting an electric current by the same process, but some are much more reluctant to do so and have to be forced!

SEMICONDUCTORS

When we get a wide variation of this type, we usually measure the ability to conduct and get some sort of idea about which materials can be considered useful as conductors of electricity, or alternatively, which can be made into useful insulators.

However, a range of materials exists between these distinct categories and two of this group have had a major impact on electronics over the past ten years, and will almost certainly do so for many years to come. They are germanium and silicon which because of their conducting abilities, are called semiconductors.

Resistivity

When measurements such as these are made it is very important to compare like with like, so that true comparisons can be made. A standard method of measuring the resistance of different materials has been adopted, and all the figures quoted refer to this same standard measurement.

A unit cube of a material is provided and the resistance in ohms is measured at 0°C between two opposite faces. Cubes of other materials are then substituted and a table of the resistivities of the different materials is built up.

Typical values of resistivities for good conductors, semiconductors and insulators are given in the following table.

Table 8.1

Material	Resistivity
Good conductors (e.g. copper, iron)	approx. 10^{-5} Ω cm
Insulators (e.g. plastics, ceramics)	approx. 10^{8} Ω cm upwards
Semiconductors (i.e. germanium, silicon)	approx. 10^{2}–10^{4} Ω cm.

THE EFFECT OF TEMPERATURE ON CONDUCTORS, INSULATORS AND SEMICONDUCTORS

As temperature changes, so do the resistivities. Certain metals have the useful property of having virtually constant resistivities. Manganin (thermal coefficient of resistivity $\pm 0.1 \times 10^{-4}$/K) and constantan (thermal coefficient of resistivity $\pm 0.4 \times 10^{-4}$/K) are typical examples. They are very useful practically for producing wire-wound resistors, since the resistances of resistors made with them will remain virtually constant regardless of the temperature. The manner of change is not the same with conductors, insulators and semiconductors, and each class of materials has to be treated individually.

Conductors

Within the structure of conducting materials the individual atoms vibrate. At zero degrees absolute or − 273°C, this vibration is absent and increases as the temperature increases. The energy of vibration comes directly from the heat energy given to the material from some outside source, hence the higher is the temperature the greater the amount of vibration. In simple terms the greater the amount of vibration the more difficult it is for the electrons to progress through the material and produce a current flow.

In general terms the increase in resistance is proportional to the temperature increase and therefore the graph showing resistance versus temperature will be a straight line in each case. Its slope will depend upon the material itself, the table below showing a list of temperature coefficients of resistivity for common materials.

Table 8.2

Copper	40×10^{-4}/K
Iron	65×10^{-4}/K
Lead	43×10^{-4}/K
Silver	40×10^{-4}/K
Tin	50×10^{-4}/K

Carbon, on the other hand, has a negative temperature coefficient and shows a negative slope when its resistivity is plotted against temperature.

Insulators

The resistivity of insulating materials generally decreases with an increase in temperature. The fall is quite dramatic although it must be remembered that the fall is from an exceedingly high value. A temperature increase of fifty per cent may reduce the resistivity from 10^{14} Ω cm to, say, 10^{10} or 10^{9} Ω cm, that is, from 100,000,000,000,000, to 1,000,000,000, a fall of many billions, but difficult to imagine. Nevertheless, the reader should appreciate that the insulation properties of an insulator are vastly reduced as temperature increases.

Humidity also affects insulating properties but its effect is really outside the scope of this book. Nevertheless, it is useful to know that as the humidity increases the resistivity falls, once again by very large amounts.

Semiconductors

The resistivity of semiconductors also falls quite rapidly as the temperature increases. The mechanism behind this reduction in resistivity will be explained fully in later paragraphs.

THE STRUCTURE OF GERMANIUM AND SILICON

The shells in the atoms perform a second useful function. They become a means of linking the atoms together into a structure. The outer electrons physically link up with those in adjacent atoms and share an orbit, which includes the nucleii of both atoms, and a sort of "glue" is formed.

Germanium and silicon both have four electrons in their outer shells. Each of these four electrons links with one other electron from a neighbouring atom, making a pair with it. In this way five atoms are linked together. Imagine a group of men with four arms. Each one stands in the crowd and shakes hands with the four people who are standing next to him. Of course, each of these four people also have a further three arms, and grasps another three people. All the people remain with their hands clasped and a link is established eventually, between all the group. Each hand clasp represents an electron pair, and each person a parent atom.

This is exactly the structure of the germanium and silicon, and it is physically very strong. From Fig. 106 you can see that each atom shares one of the electrons in its outer shell with each of its four neighbours, and that by doing so each atom effectively has eight electrons in its outer shell. As we have previously seen, materials whose atoms have full outer shells make good insulators. The electron bonds are known as pair bonds, and form

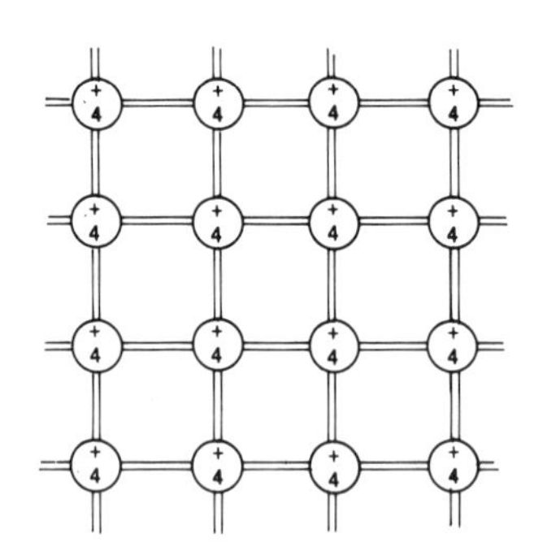

Fig. 106. *Simple covalent bonding*

what is termed a crystal lattice structure. Figure 106 shows a simplified two dimensional representation of the crystal structure. This type of bonding is termed *covalent* bonding.

At this point we must note that the atoms still have the correct balance of charges, with as many electrons as protons, and still behave as neutral overall. This fact is very important later in this chapter and should be remembered.

INTRINSIC SEMICONDUCTION

When the temperature of a body rises, heat energy has been given to it. Part of this energy causes the individual atom to vibrate. The more heat energy is given, the more violent the vibration. The silicon and germanium atomic structure, when subjected to normal room termperatures, has a sufficiently violent vibration to cause some of the pair bonds to break. The electrons are then free in the structure and are available to become part of a current if needed. It is most important to consider what occurs in the individual atoms when this pair bond is fractured. The bond breaks and the electron is free to move away. The neutral charge balance is now disturbed. The nucleus has one more positive proton than there are negative electrons in orbit. The place where the electron was is now lacking a negative charge. It is sometimes referred to as a void or hole. However, another electron may be able to move and fill that hole, but in doing so, it will have left another hole somewhere else in the structure. We can deduce two facts from this. The hole behaves as a positive charge and will change its position within the structure in the opposite direction to the movement of the electrons. Imagine a row of cinema seats that are full, except for the end one. All the people are asked in turn to move up one seat. The people, who represent the electrons, shift one seat along to the left, but the empty seat, the hole, moves across the row to the right. It finishes up a full row away, although the people have only moved one seat each. The hole, then, can be thought of as being a carrier of current just like the electrons, but in this case positive charges are moving about.

This idea of two currents carriers has to be remembered when dealing with the principles of semiconductors.

An e.m.f. applied across the crystal will result in a current through it, but the numbers of fractured bonds are relatively few, and the currents will be small. Nevertheless, conduction does take place and, because it is much smaller than that which good conductors would give for a given e.m.f., and yet much larger than for a true insulator, the material is classified as a semiconductor. The semiconduction which occurs because of the temperature is called intrinsic semiconduction, and currents of the order of only one or two microamps are typical.

DOPED SEMICONDUCTORS

The size of the current will depend upon how many carriers are available in a given volume of material. In general, higher intrinsic currents will flow in germanium than silicon.

Consider what might happen, if instead of relying on the fracture of pair bonds to provide all the current carriers we need, we could sprinkle a few in amongst the crystal structure to help things along.

Here nature helps a great deal. It is convenient that electrons taken from one element are the same as those from another, as it was found that it is quite practical to introduce the extra carriers in the form of electrons from a different material. The process is called doping and is widely adopted to produce the correct type of semiconductor from pure silicon or germanium.

n-type Doping

Let us think again about the four-armed men. All of them, you remember, have clasped another four hands and are holding on tightly. Now, imagine that amongst them one man is introduced with five arms. Four of his hands will each grasp those of a neighbour, but one hand will be free. This is just what happens when we introduce an atom which has five electrons in its outer shell.

Four of the electrons in the outer shell of the atom will pair with four electrons from neighbouring atoms to give that shell eight electrons, while its fifth electron will have to form its own shell. As you will remember, this is similar to the arrangement of electrons in a metal, where the outer shell electron is very weakly held and is thus free to break off and act as a current carrier. It thus becomes a free electron and will move as a current carrier in a particular direction if an external influence such as an e.m.f. is applied. Remember, when it moves away it leaves a positive hole which will also shift around within the structure. The spare charge introduced in this case is negative and the material is known as n-type semi-conductor. Typical doping agents are arsenic, phosphorus and antimony. The amounts involved vary dependent on the intended use of the semiconductor, but are

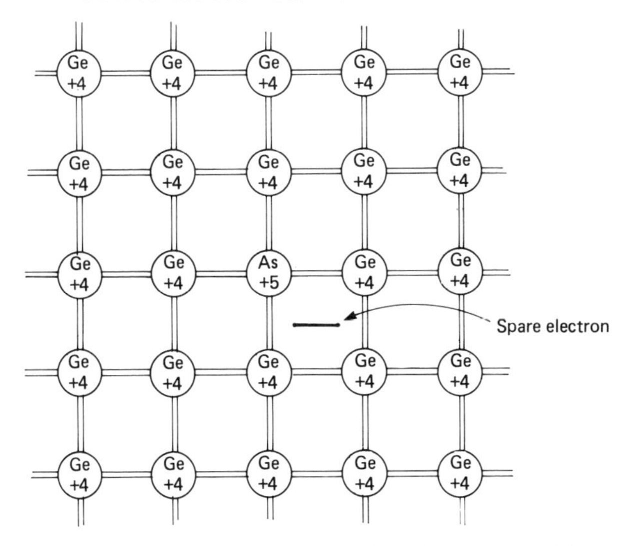

Fig. 107. *Structure of a typical n-type semiconductor material (germanium doped with arsenic)*

usually about one part in a million. Figure 107 illustrates this type of doping.

p-type Doped Semi-conductors

An alternative method of doping, using similar principles, is to introduce a doping impurity having only three electrons which can be bonded. In this case the doped material finishes up with places in the structure which are lacking electrons. You will remember that we have called these holes and have assumed that they are positively charged, hence the name p-type semiconductor. Figure 108 shows the structure of a typical p-type semiconductor.

Both types of doped semiconductor have very wide application these days and can be based on germanium or silicon. Silicon, however, is proving to be the more versatile, and the newest technologies are based on silicon. Typical p-type doping agents are aluminium, gallium and indium.

THE pn JUNCTION

A piece of doped silicon or germanium has little use, other than as a means of circuit resistance. However, major developments

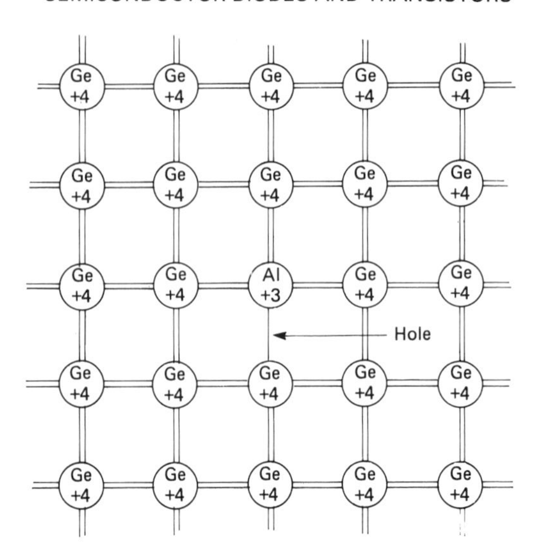

Fig. 108. *Structure of a typical p-type semiconductor material (germanium doped with aluminium)*

have taken place using the behaviour of a junction formed between a piece of n-type and a piece of p-type material. It is this behaviour which we must now study in more detail.

Imagine that we have two such pieces of material, and that they are physically brought together to form a junction. You will recall that one piece, the n-type semiconductor, has spare electrons free to move about. You will also remember that the other piece, the p-type semiconductor, has positive "holes" which are also effectively moving (but remember that over all the materials are neutrally charged).

Near the junction some of the free electrons will move across the junction and fill some of the holes. These free electrons are said to *neutralise* the holes. They do this obeying the normal rule of charges, that positive and negative charges attract. Let us look at one example of this happening.

As shown in Fig. 109, electrons cross the junction from the n-type material, filling the holes in the structure of the p-type material and establishing pair bonds. This seems simple enough, but two very important by-products of this process must be considered. When an electron leaves its parent doping atom in the n-type material, it disturbs the charge balance. The parent atom now has a positive charge. Conversely, when that same

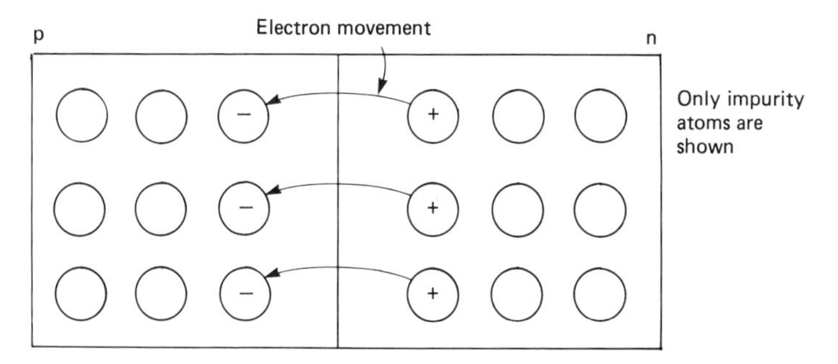

Fig. 109. *The buildup of potential at a pn junction*

electron arrives at an impurity atom in the p-type material, it disturbs the charge balance there also, and that atom now has an extra negative charge.

The Depletion Layer

The above process does not happen once, but many times close to the junction. Each time the charge imbalance is further affected. Eventually a situation will be reached that on the p side of the junction, a sizeable negative charge will have built up. An electron which has a negative charge and now tries to get across, will be repelled by the like charge, and will not be able to cross.

Eventually, a state of equilibrium will be reached. All the charge carriers close to the junction have recombined and in doing so produced a potential difference across the junction itself. Electrons now attempting to reach the p-type holes, do not have enough energy to penetrate the negative charge which exists on the p-side of the junction, and are repelled. A similar situation exists in the p-type material. The region very close to the junction is thus depleted of charge carriers and is called the *depletion layer*.

We have already explained that a minority of charge carriers exist in the materials because of thermal excitement of the atoms. These minority carriers will be attracted by the potentials set up in the depletion layer and will be swept easily across the junction. The final state of equilibrium is therefore when this minority carrier current is equalled by that number of majority carriers which have enough energy to cross the barrier. The junction will remain in this state until some different conditions are introduced. The actual potentials involved at the junction are about 0.3 V for germanium and 0.7 V for silicon.

The Useful Property of the pn Junction

When an external potential is applied, it is found that current can be driven across the junction in one direction, but not in the other. The device is uni-directional. We have already seen in an earlier chapter that this is the exact property needed in a rectifier. The pn junction is thus a semiconductor rectifier and has very many applications in modern electronics, e.g. in power circuits, rectifiers, demodulators, detectors, damping circuits, etc.

The Forward Biased Junction

Briefly we can examine the above effect. When an external battery is applied with the polarity shown in Fig. 110 the junction

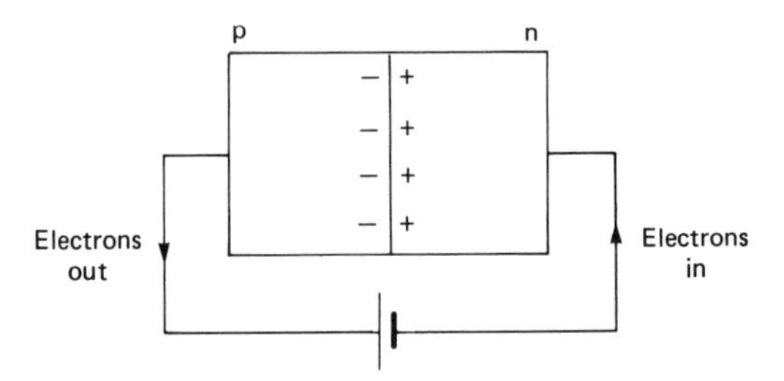

Fig. 110. *The forward biased pn junction*

is forward biased. Holes in the p-type material are repelled by the battery potential, as are electrons in the n-type. These carriers then have sufficient energy to cross the junction barrier potential easily. They are so much more abundant than the minority carriers which are moving in the opposite direction, that once a significant current flow is established the resistance of the diode falls and the current flow across the junction increases rapidly with increasing voltage. The relationship between the diode forward current and the applied voltage is shown in Fig. 111.

The Reverse Biased Junction

When a battery of reverse bias polarity is connected the effect is to draw majority carriers away from the junction. This emphasises the depletion layer and the existing small numbers of carriers which manage to cross is reduced still further. However, the flow of minority carriers across the junction is increased. The equilibrium is disturbed and the battery has to provide a small current even when the diode is reverse biased.

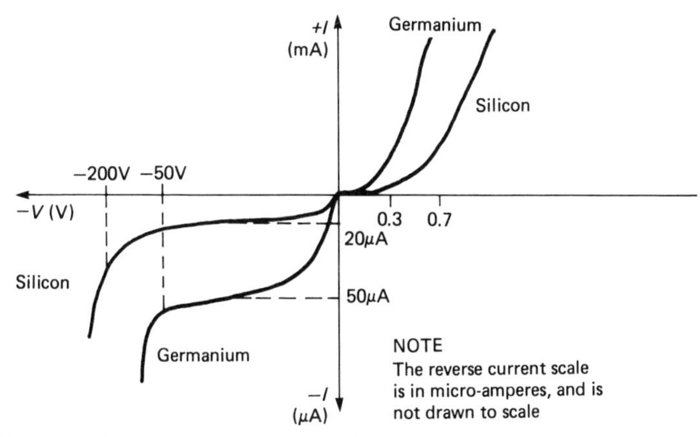

Fig. 111. *The forward and reverse characteristics of semiconductor diodes.* NOTE *The reverse current scale is in microamperes and the forward current is in milliamperes*

The size of this current will depend on the temperature at which the device is operating. Typical currents for normal room temperatures are about 0.02 μA for silicon and 1.0 μA for germanium. This is known as the reverse saturation current I_0. The shape of the reverse characteristic is also given in Fig. 111, and it should be noted that the scales used for forward and reverse currents are very different.

The British Standard Symbol for pn junction
The British Standard symbol for a semiconductor diode is shown in Fig. 112.

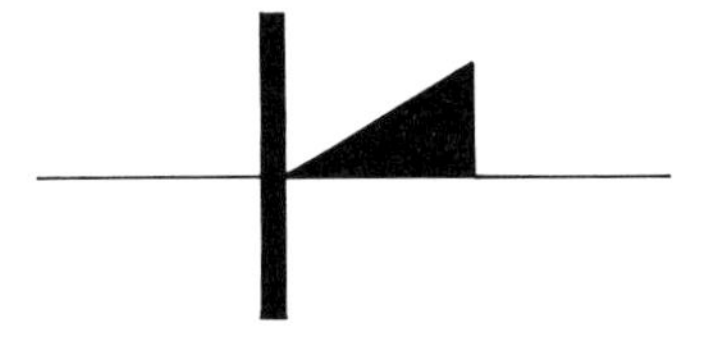

Fig. 112. *The British Standard symbol for the semiconductor diode*

THE BIPOLAR TRANSISTOR

The transistor is a three-terminal device which may be used in many different ways for switching or amplifying circuits. There are many types of transistor but the bipolar (also known as the bipolar junction transistor) is the most common.

It is made up of three layers of doped semiconductor material. When two n-layers sandwich a p-layer it is known as an npn transistor. Conversely, when n-type material is between two p-layers, a pnp transistor is formed. The three layers are called emitter (E), base (B) and collector (C). Each has connections made to it via connecting wires or terminals and the whole is usually encapsulated is some form of protective case. Several methods of production are used and, for individual transistors, alloying is one common technique. In this method, pure silicon is

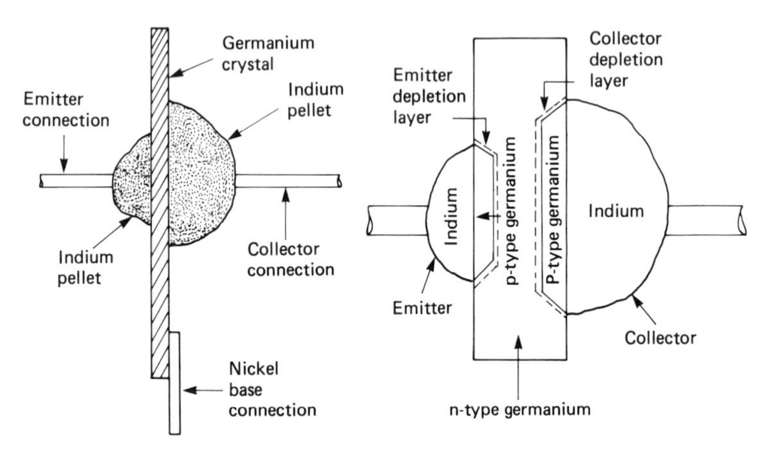

Fig. 113. *The make-up of the alloy type of transistor*

doped with a suitable impurity to produce n-type semiconducting material. A thin wafer, about 0.02 mm thick is then taken and small pellets of indium placed on opposite surfaces. The pellets are diffused into the surface by heat treatment, using temperatures upwards of 550°C, to a carefully controlled depth. Connecting leads are then attached and the whole assembly mounted in a sealed container (*see* Fig. 113).

More modern production builds up the materials using a layering technique on a suitable silicon base or *substrate*. The techniques of manufacture are continually evolving to satisfy the demand of power handling capacity, size, frequency, characteristics, etc. The transistor action itself is identical in each case.

British Standard symbols for bipolar transistors

The standard symbols for transistors are shown in Fig. 114. The arrow indicates "hole flow" and is therefore equivalent to con-

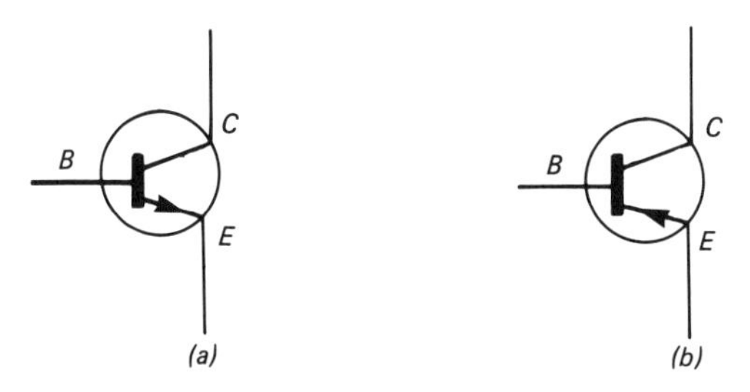

Fig. 114. *The British Standard symbols for:* (a) *npn; and* (b) *pnp transistors*

ventional current flow. It is the means of distinguishing between the pnp and npn symbols.

THE BASIC TRANSISTOR ACTION

In effect the transistor is two pn junctions connected back to back. When we consider the basic action, the two junctions are biased differently, one forward and one reverse. The external d.c. supplies are first connected so that the internal currents can flow. The "emitter to base" junction is forward biased, and the "base to collector" junction is reverse biased. A simple aid to memory here is that "positive to p" is forward bias.

The npn action

The emitter generates a large number of majority carriers which can be swept across the forward biased emitter-base junction quite easily. In the npn transistor electrons flow readily across this junction into the base region. Now, although the base-collector junction is reverse biased to the majority carriers, the electrons, which now find themselves in the p-type base material, are in the same condition as the minority carriers which already exist in the p-region (they are electrons in a p-type material). They can therefore transfer through the base region and across the second junction, which to the minority carriers is also forward biased, to be gathered by the collector.

While passing through the p-type base material, a few electrons recombine with holes which exist there and are lost to the emitter-collector current I_{EC}. This means that the base material will steadily build up extra negative charges disturbing the relative bias potentials. In order to maintain the bias potentials steady, a small flow of electrons is taken from the base, generating as many positive holes in the base material as there are extra negative electrons being captured.

The total flow of electrons is illustrated in Fig. 115. Note that one additional current is shown. This is a minority carrier current which flows across the reverse biased base-collector junction and will depend upon the operating temperature. It is often referred to as a leakage current, and is labelled I_{CBO} to distinguish it from majority carrier currents. To be precise, the base current will therefore consist of two components in opposite directions.

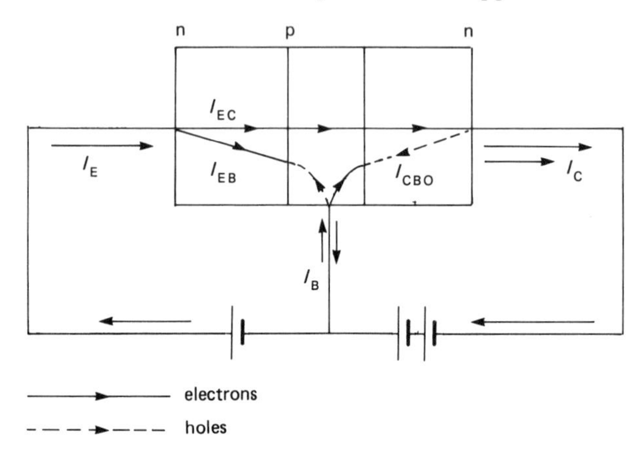

Fig. 115. *Electron and hole migration in an npn transistor*

However, the leakage current is small compared to the base current, which itself will only be a few microamps. The collector current I_{EC} will be several milliamps and will be fractionally smaller than the emitter current, usually by about one per cent due to the base current. Although it does exist, the leakage current is very small in comparison with the collector current and contributes little. We can say that $I_E = I_B + I_C$, which holds good for our circuit conditions.

pnp Transistor Currents

When a pnp transistor is used the basic action is identical to that of an npn transistor, except that to set up the correct biasing condition, the polarities of the supplies must be changed around. All previous electron flow can be replaced by hole flow, remembering that only electron carriers can move in the external connections. Electrons which are attracted from the emitter by the positive terminal of the bias supply, generate holes in the p-type material which can then migrate through it. The total current flow is shown in Fig. 116.

Once again a leakage current will occur, its size dependent on the operating temperature. In this case, the minority electrons in the collector will migrate to the base-collector junction and will recombine with holes which are being generated in the base material. Again currents of only about 1 μA for germanium and about 0.02 μA for silicon are involved.

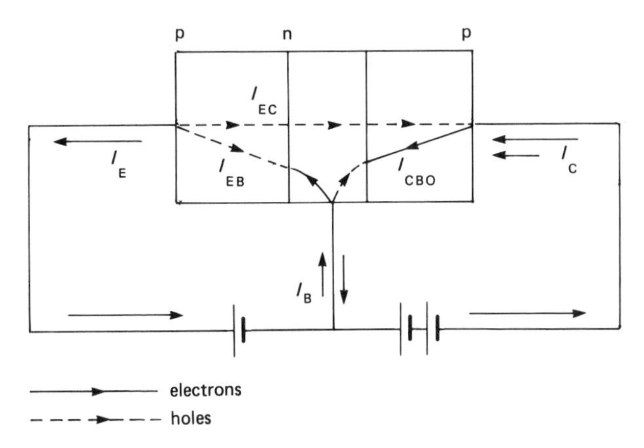

Fig. 116. *Electron and hole migration in a pnp transistor*

Static Operating Conditions

It must be remembered that the currents which we have been considering are those brought about by the application of d.c. bias potentials. They are not those under which the transistor will operate when a signal is being amplified, although the basic operation just described is fundamental to the amplifying action.

TRANSISTOR CONFIGURATIONS

The transistor is a three-terminal device and, because there is always one terminal common to the input and output circuits, there are three possible connection patterns which can be used. They are usually called the transistor configurations and are listed as:

(*a*) common base mode;
(*b*) common emitter mode; and
(*c*) common collector mode.

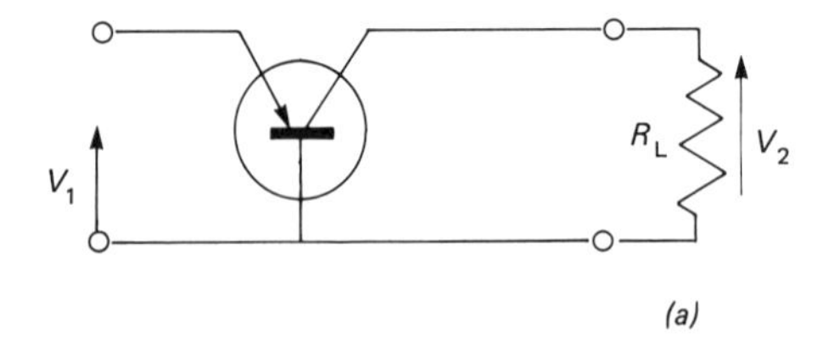

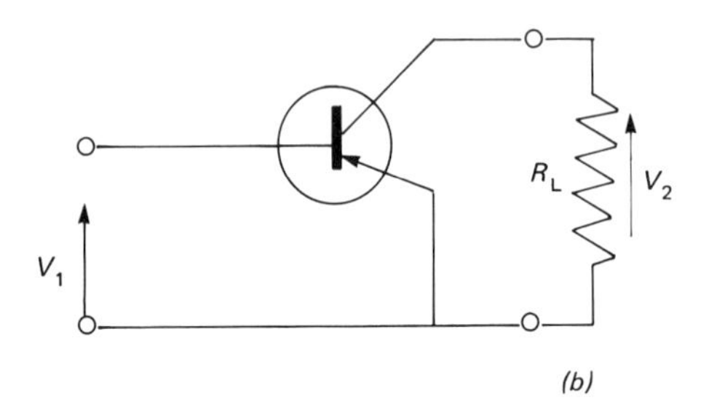

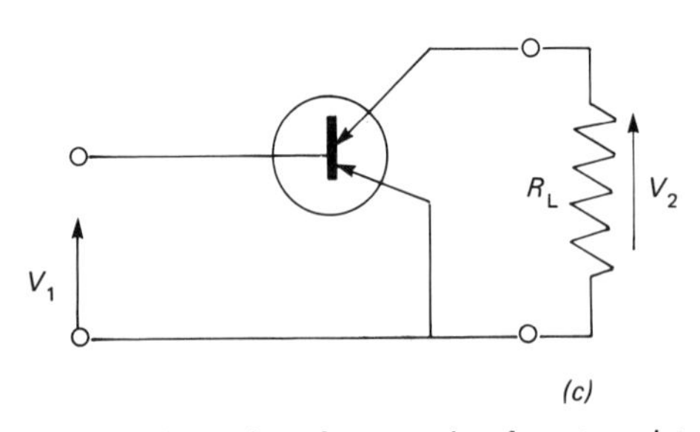

Fig. 117. *The three basic modes of connection for a transistor:* (a) *common base mode;* (b) *common emitter mode; and* (c) *common collector mode*

Each mode of operation sets up different operating conditions and the circuit designer uses the one which is most suitable for his needs. In this unit it is important that you understand the three basic methods of connection. Figure 117 shows the three modes of connection. The modes (*a*) and (*b*) are the most commonly used in practice; (*c*) being used for somewhat specialised purposes, but it is nevertheless useful where it meets the circuit requirements.

The common collector mode is often referred to as an *emitter follower* circuit.

The Correct Potentials

When a practical circuit is being set up, it is not sufficient to say that the various junctions must be forward or reverse biased. The actual potentials must be stated. The current/voltage graph for the junction diode (*see* Fig. 111) shows that little current flows until the emitter-base voltage exceeds 0.7 V for silicon and 0.3 V for germanium. It is necessary for these voltages to be exceeded so that the emitter-base current is of a reasonably high value. Typically, nine or ten volts are connected at the collector although the circuit will operate with considerably lower voltages. A typical voltage distribution is shown for npn transistors in Fig. 118.

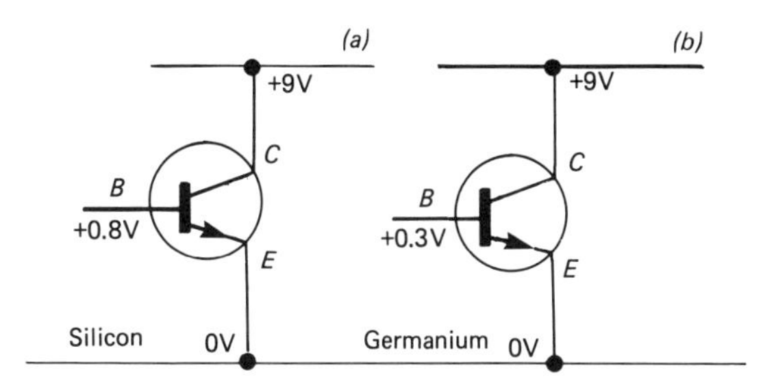

Fig. 118. *Voltage distribution for* (a) *silicon; and* (b) *germanium biasing*

Static Characteristics

Slight changes in these potentials may well have considerable effects on the performance of the transistor. Alternatively, very little effect may be obvious, e.g. when the collector potential is reduced, say, from 9 V to 8 V. In order to gauge these changes the transistor is built into a simple circuit and the appropriate bias conditions applied. The circuit is constructed so that the applied potentials can be changed while measuring the changes in the current flowing in various parts of the device. Typical circuits are shown in Fig. 119. Since there are three terminals the potentials applied to one pair must be held constant whilst that to the other is varied and the appropriate current measured. The change in circuit currents are then noted, noting which values were held constant.

In this way a complete picture of the behaviour of the transistor under different operating conditions is built up. The transistor operating mode can then be changed and the measurements repeated, showing how the performance of the transistor depends upon circuit connections.

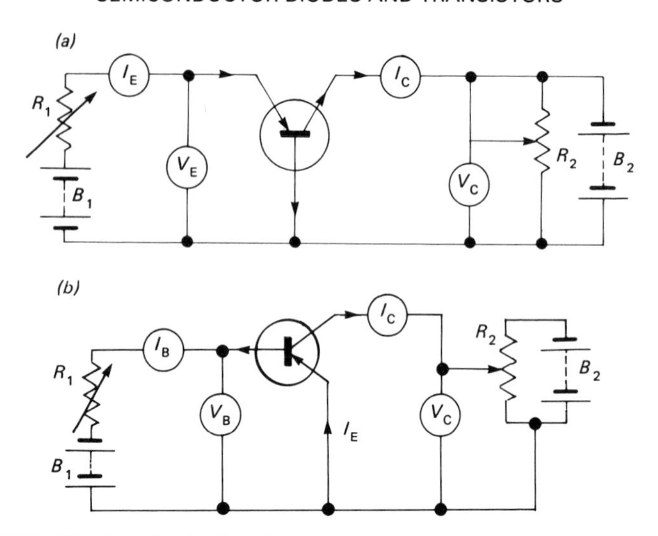

Fig. 119. *Circuit suitable for obtaining the:* (a) *common base characteristics; and* (b) *common emitter characteristics*

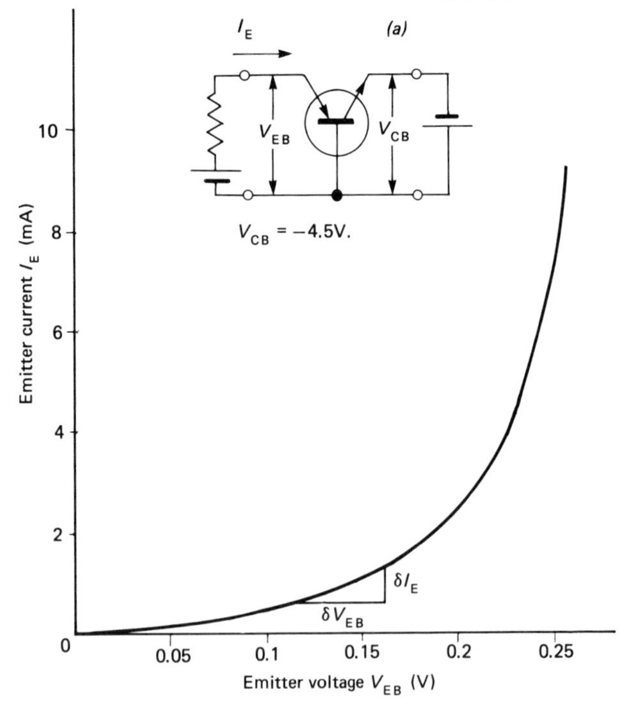

Fig. 120. *Input characteristics:* (a) *common base mode;* (b) *common emitter mode*

The graphs which may be drawn from the figures obtained are known as the static curves for the device. Typical examples are shown in Figs. 120 and 121.

Input Curves

These show how the basic transistor input current varies as the bias across the first junction is changed.

In each case it is seen that the collector voltage is held constant whilst the respective emitter or base currents are measured for various values of bias voltage (*see* Fig. 120).

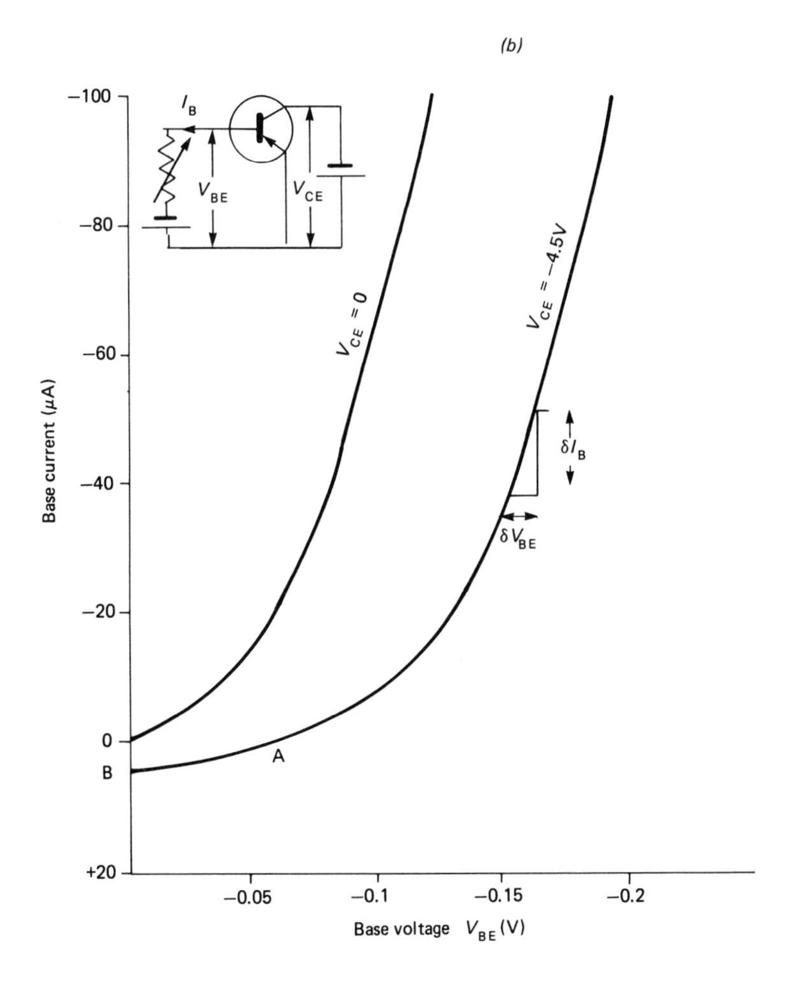

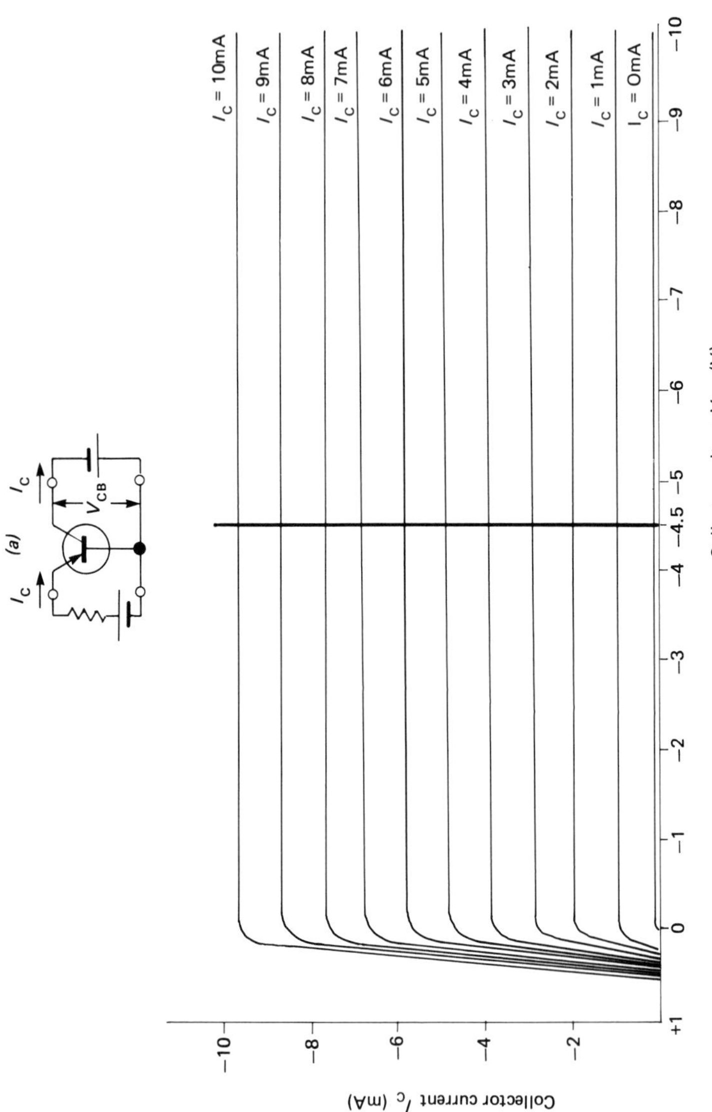
(a)
I C
V CB
I C = 10mA
I C = 9mA
I C = 8mA
I C = 7mA
I C = 6mA
I C = 5mA
I C = 4mA
I C = 3mA
I C = 2mA
I C = 1mA
I C = 0mA
+1 0 −1 −2 −3 −4 −4.5 −5 −6 −7 −8 −9 −10
Collector voltage V CB (V)
−10 −8 −6 −4 −2
Collector current I C (mA)

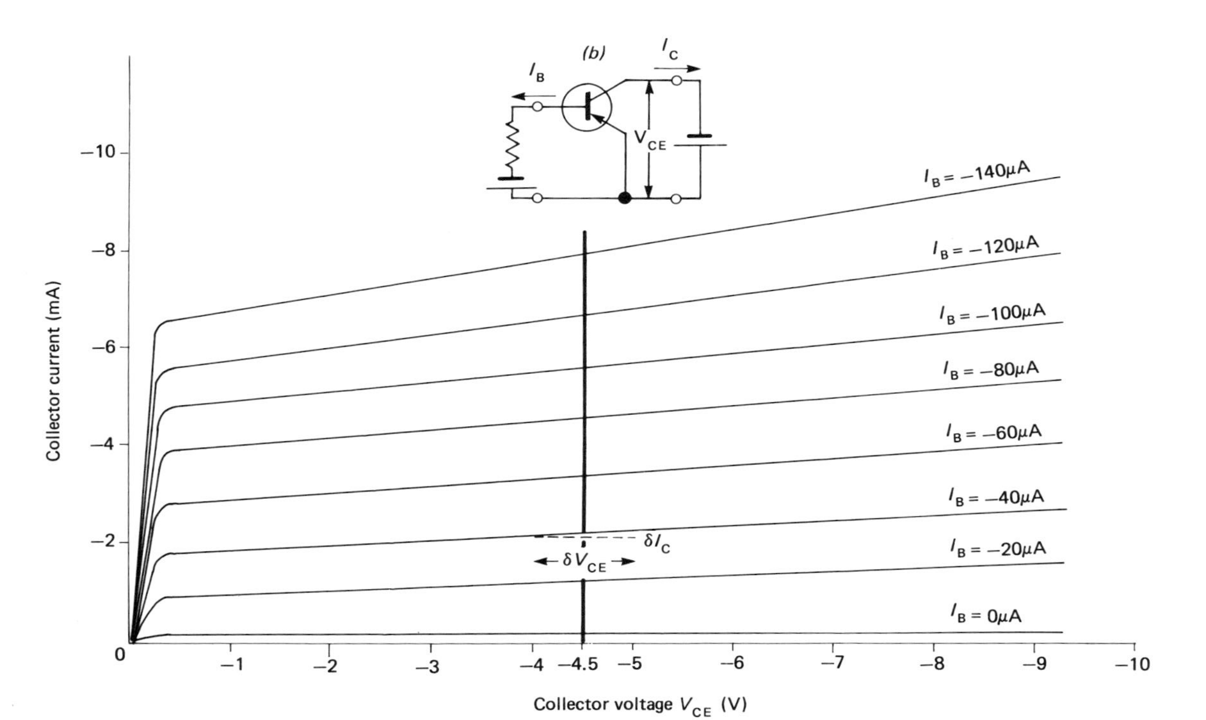

Fig. 121. *Output characteristics:* (a) *common base mode;* (b) *common emitter mode*

Output Characteristics

The output characteristics show how the output collector current is affected by changing the applied voltage to the collector, while either the emitter or base currents remain constant depending on the operating mode (*see* Fig. 121).

Transfer Characteristics

A third set of static characteristics are generated when the relationship between input and output currents is investigated. These are known as the transfer characteristics and examples are shown in Fig. 122.

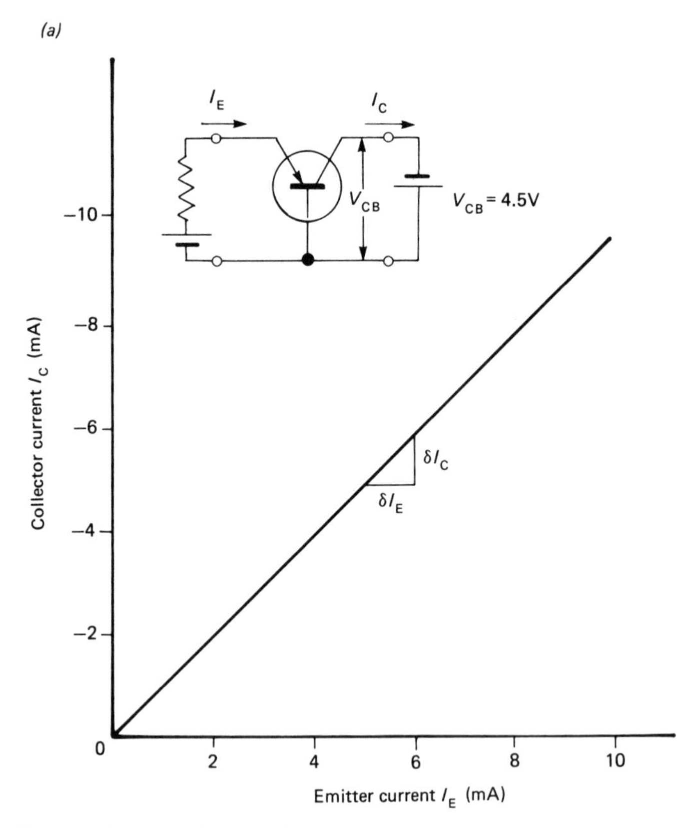

Fig. 122. *Transfer characteristics:* (a) *common base mode;* (b) *common emitter mode*

SELF-ASSESSMENT QUESTIONS

1. The basic make-up of an atom is:
   (*a*) central electrons around which orbit neutrons;
   (*b*) a central nucleus around which orbit electrons;
   (*c*) a central group of protons around which orbits the nucleus;
   (*d*) a central nucleus around which orbits a group of neutrons.
2. The respective charges on the particles within the atom are:
   (*a*) protons positive, electrons negative, neutrons neutral charge;

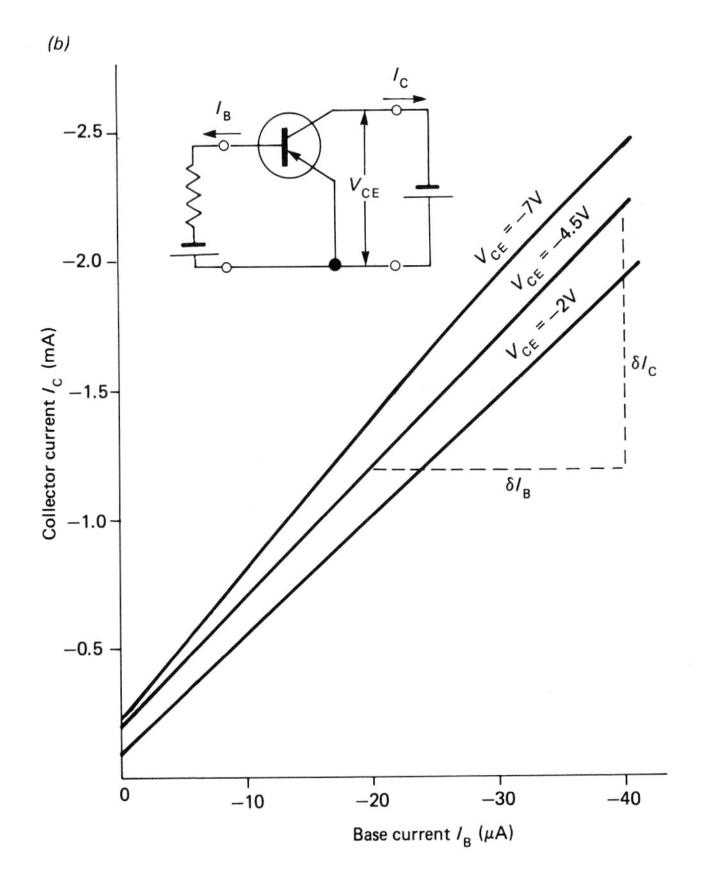

(*b*) protons no charge, electrons negative, neutrons neutral charge;
(*c*) neutrons positive, electrons negative, protons no charge;
(*d*) electrons positive, neutrons no charge, protons negative.

3. A complete atom exhibits the following charge:
(*a*) no charge at any time;
(*b*) positive at all times;
(*c*) negative at all times;
(*d*) positive only for short periods of time.

4. When an increase in temperature in a pure crystal of silicon gives rise to the generation of carriers so that a current can be made to flow through the material, this is known as:
(*a*) extrinsic semiconduction;
(*b*) p-type aneptor semiconduction;
(*c*) intrinsic semiconduction;
(*d*) n-type donor semiconduction.

5. Describe carefully how impurity addition to pure germanium or silicon can be made to produce semiconductor material.

6. Describe specifically how the choice of impurity gives rise to p or n type materials.

7. Describe the process of migration which takes place when a piece of n-type semiconductor is joined with a piece of p-type material.

8. Describe how this pn junction allows current flow in one direction only.

9. Explain carefully the terms:
(*a*) majority carriers;
(*b*) minority carriers.

10. Electrons can be described as:
(*a*) minority carriers in p-type and minority carriers in n-type semiconductors;
(*b*) majority carriers in p-type and minority carriers in n-type semiconductors;
(*c*) majority carriers in n-type and minority carriers in p-type semiconductors;
(*d*) majority carriers in n-type and majority carriers in p-type semiconductors.

11. Show how a bipolar transistor can be produced using a pnp sandwich of semiconductor material and describe its fundamental action.

12. Explain what the term (*a*) common emitter; (*b*) common

collector; and (*c*) common base mean in terms of methods of transistor connection.

CUMULATIVE QUESTIONS, CHAPTERS 5–8

1. A 50 V sinusoidal voltage has a maximum value of 300 V. How long after passing through zero while rising positively will the voltage attain a value of 215 V?

2. An e.m.f. of a sinusoidal wave form has a maximum value of 354 V, calculate:
   (*a*) the r.m.s. voltage;
   (*b*) the average voltage;
   (*c*) the form factor.

3. A current in a circuit can be expressed as $I = 141.4 \sin 377t$, calculate.
   (*a*) the r.m.s. current;
   (*b*) the frequency; and
   (*c*) the value of I when $t = 0.0035$s.

4. The following values describe a half cycle of alternating current.

Angle	0	20	40	60	80	100	120	140	160	180
Current	0	15	24	35	54	68	64	35	12	0

Draw a smooth curve through these points and using the mid-ordinate rule calculate:
   (*a*) the average value; and
   (*b*) the r.m.s. value.

5. The current in the two paths of a parallel circuit are given by:

$$I_1 = 40 \sin (314t - \pi/4); \text{ and}$$
$$I_2 = 20 \sin (314t + \pi/6).$$

Draw phasors representing these currents and add them to find the total current. Express the answer as an r.m.s. value.

6. At what supply frequency will a coil of inductance 0.075 H have an inductive reactance of 23.55 Ω?

7. A circuit has an impedance of 80 Ω. If the supply voltage is 240 V at 50 Hz, calculate the value of current.

8. A non-inductive resistor is connected in series with a pure inductance. If the p.d. across the inductance is 192 V and the supply voltage is 240 V, calculate the p.d. across the resistor.

9. If an inductor has an impedance of 31 Ω at a supply frequency of 50 Hz when the value of inductance is 0.07 H, what is its resistance.

10. A series a.c. circuit dissipates a power of 2.88 kW. If the supply voltage is 240 V and the current lags the voltage by 53°, calculate:
   (*a*) the power factor;
   (*b*) the current;
   (*c*) the impedance, inductive reactance and resistance.

11. An inductor has a resistance of 6 Ω and an inductive reactance of 3.5 Ω. If the a.c. supply voltage is 30 V, calculate:
   (*a*) the impedance;
   (*b*) the current;
   (*c*) the power factor;
   (*d*) the power.

12. A coil has an inductance of 0.2 H and negligible resistance. Calculate the current when the coil is connected to a.c. supplies of 100 V at:
   (*a*) 30 Hz;
   (*b*) 500 Hz.

13. A coil which has a resistance of 20 Ω and an inductance of 0.15 H is connected in series with a 100 μF capacitor across a 230 V, 50 Hz supply, calculate:
   (*a*) the voltage across the coil;
   (*b*) the power factor of the circuit.

Sketch the phasor diagram for this circuit.

14. A voltage given by 100 sin 314t is applied to a circuit comprising a half-wave rectifier in series with a resistor of 50 Ω. Ignoring any leakage resistance in the rectifier, calculate the average and r.m.s. values of the current.

15. Two sinusoidal voltages with peak values 50 V and 20 V respectively have a phase difference of 30°. Draw the phasor diagram and then calculate the peak, r.m.s. and average values of the resultant voltage.

16. A simple potentiometer is calibrated by using a standard cell of e.m.f. 1.433 V, a null point being found at a point 60 cm from the common end of the slide wire.

When the standard cell is replaced by a cell of unknown voltage a new null point is found 50 cm from the common end. Calculate the e.m.f. of the second cell.

17. A moving-coil galvanometer gives a full-scale deflection when a current of 5 mA is passed through it. If the resistance of the coil is 7.5 Ω, calculate the resistance of the shunt required to enable the meter to measure currents of up to 0.5 A.

18. An unknown resistance is measured by using a Wheatstone bridge. When balanced, the arms AB and BC have resistances of 1,000 Ω and 10 Ω respectively and the

variable resistance AD is set at 2 Ω. Calculate the value of resistance CD.

19. A sinusoidal alternating current has a maximum value of 10 A. If this current is then rectified full wave, calculate the readings which would be obtained on an a.c. ammeter and a moving-coil ammeter which are connected in series to measure the current.

20. If a rectifier type voltmeter is calibrated to accurately measure the r.m.s. value of a sinusoidal voltage, calculate the correction factor which must be applied in order for the instrument to correctly indicate the r.m.s. values of:

(*a*) a square wave voltage; and
(*b*) a voltage having a form factor of 1.15.

21. Describe how an increase in temperature effects the resistivities of (*a*) conductors, (*b*) semiconductors and (*c*) insulators.

22. Describe what the static characteristics of a bipolar transistor are. Give examples showing typical input, transfer and output characteristics.

Appendix

ANSWERS TO SELF-ASSESSMENT AND CUMULATIVE QUESTIONS

Chapter 1

1. 0.32 A 2. 6,000 μF 3. 0.00468 C 4. (*b*) 5. (*c*) 6. (*a*) 8 V; 12 V; 24 V; 1 A; (*b*) 44 V; 5.5 A; 3.67 A; 1.83 A 7. 0.38 A 8. $\frac{11}{19}$ = 0.58 A

Chapter 2

1. 1,400 At 2. 1.6 A 3. 1.4 Wb 4. (*b*) 5. 702.5 At/m 6. (*a*) 7. 0.754 A 8. 3.18 $\times$ 10^6 At/Wb; 2.83 $\times$ 10^{-6} Wb 9. Cast iron has a low saturation density, low permeability and high hysteresis loss. Cast steel has a high saturation density, high permeability and low hysteresis loss. Armature laminations have a high saturation density, high permeability and low hysteresis loss. Cast iron.

Chapter 3

1. (*d*) 2. (*b*) 3. 200 μF; 0.02 C 4. (*a*) 6.7 $\times$ 10^6 V/m; (*b*) 2.03 $\times$ 10^4 V 5. (*a*) 590 pF; (*b*) 0.295 μC; (*c*) 166.67 kV/m; (*d*) 7.375 $\times$ 10^6 C/m^2 6. 6 μF; 6 $\times$ 10^{-3} C; 6000 V; 300 V; 100 V 7. 33.9 cm^2 8. (*a*) 1.195 μF; 4.25 μF; (b) 0.933 μF; (*c*) 591.75 V; 158.25 V 9. 1,875 mA; 150 V; 375 V

Chapter 4

1. 10.8 V 2. 26.7 V; 2.67 A; 22.7 N 3. 1.92 mH 4. 1.5 Wb 5. 2 mH 6. 1.8 J 7. 5 A 8. 80 V 9. 4.5 H 10. 75 V; 60 V 11. 750

Cumulative Questions, Chapters 1–4

1. (*d*) 2. (*a*) 3. 1.8 A; 1.2 A; 81 V; 243 W 4. 2.03 A; 1.24 A; 0.79 A 5. (*a*) 2.94 A; (*b*) 1.71 A; 1.23 A 6. 1.89 A; 0.26 A; 1.63 A 7. 47.89 mA; 70.6 mA; 22.71 mA 8. 1.21 A; 0.27 A; 0.94 A 9. 1,500 10. 1.5 mWb 11. 2.55 μWb 12. 4.6 mWb 13. 3.9 A 14. 2 mH 15. 500 μC; 250 V; 166.7 V; 83.3 V; 0.0208 J 16. 6.06 μF 17. 3.75 μF; 1,500 μC 18. 1.66 μF; 12 μF 19. 12 μF 20. 15 μF 21. 6 μF; 120 μC 22. 0.0531 μF 23. 200 V; 1,200 μC; 2,000 μC 24. 1.416 mm 25. 1.92 mH 26. 1.5 mWb 27. 7.2 J 28. 1.25 H 29. 5A

30. 450 W 31. (*a*) 100; (*b*) 12 V 32. 31.3 A 33. 16.7 A 34. (*a*) 833; (*b*) 6.25 A 35. 69.2 V 36. 236 V; 308 A

Chapter 5

1. (*c*) 2. 31 V; 42.4 V 3. 2 ms; 5.8 A 4. 15.5 sin (ωt – 10.245) 9. (*a*) 100 V; (*b*) 70.7 V; (*c*) 400 Hz; (*d*) 1/400 s 10. Form factor = r.m.s. value/average value; and defines the shape of the wave.

Chapter 6

1. (*c*) 2. (*c*) 3. (*d*) 4. (*c*) 5. (*a*) 9. (*d*) 10. (*b*) 11. 636.5 Ω; 0.17 A 12. 10.75 A; 57°30′ 13. 5.32 A; 57°46′; 567 W

Chapter 7

1. (*c*) 3. Resistor of resistance 0.051 Ω in parallel with instrument 4. 50.9 mA 5. 40 Ω 6. (*a*) 1.67 Ω in parallel; (*b*) 9,985 Ω in series 7. 19,996 Ω in series; tappings at 9,996 Ω and 3,329 Ω 8. (*a*) moving-coil galvanometer; (*b*) moving-iron galvanometer; (*c*) rectified moving-coil galvanometer.

Chapter 8

1. (*b*) 2. (*a*) 3. (*a*) 4. (*c*) 10. (*c*)

Cumulative Questions, Chapters 5–8

1. 2.55 ms 2. 250 V; 225.18 V; 1.11 3. 100 A; 60 Hz; 127.8 A 4. 33.44A; 39.9 A 5. 34.8 A 6. 50 Hz 7. 3A 8. 144 V 9. 22 Ω 10. 0.6; 20 A; 12 Ω; 7.2 Ω; 9.6 Ω 11. 7.5 Ω; 4 A; 0.8; 96 W 12. 2.65 A; 0.159 A 13. 468 A; 0.795 lagging 14. 0.637 A; 1.0 A 15. 68 V; 48.1 V; 0 V 16. 1.1942 V 17. 0.07575 Ω 18. 0.02 Ω 19. 7.07 A; 6.36 A 20. (*a*) 0.9 V; (*b*) 1.036 V

Index

alternating current circuits, single phase, 118–37
alternating quantities, 94
  amplitude, 98, 113–14
  average value, 101
  cycle, 100
  frequency, 100, 113–14
  instantaneous value, 101
  peak value, 98
  root-mean-square value, 102
  waveforms, 97
ampere, 1
atomic structure, 166
average power, 124, 131, 136

bipolar transistor, 173
  British Standard symbols, 174–5

calibration errors, instruments, 147
capacitive reactance, 122
  variation with frequency, 125
capacitor charge and discharge curves, 46
capacitors, 43
  charge, 45
  multi-plate capacitor, 52–3
  parallel connection, 55
  parallel-plate capacitor, 50
  series connection, 54
  series-parallel connection, 56
capacitor types, 59
  air dielectric, 59
  ceramic, 61
  electrolytic, 61
  mica, 60
  polyester, 60
  waxed paper/foil, 59
cathode ray oscilloscope, 149
circuits, alternating current, 119
circuits, direct current, 7–9
  parallel circuits, 8, 9
  series circuits, 7, 9
coercive force, 39
coercivity, 39
conductors, good and bad, 162
control torque, instruments, 141, 145
current growth in inductive circuit, 80

damping torque, 141
depletion layer, 171
dielectric loss, 57
dielectrics, 48
digital meters, 158
doping, n- and p-type, 168–9

eddy current damping, 143
electrical energy, 2
electric charge, 43, 44
electric field strength, 47
electric flux density, 47, 48
electromagnetic units, 27–31
  magnetic flux, 29
  magnetic flux density, 29
  magnetising force (magnetic field strength), 28
  magneto-motive force, 27
  permeability, absolute, 29
  permeability of free space, 30
  permeability, relative, 30, 34
  permeability, variation, 34
  reluctance, 31, 34
electromagnetism, 27
electromotive force, 2, 82
emf induced in a coil, 73
extension of instrument range, 148–9

Faraday's discoveries, 68, 79
Faraday's laws, 70
ferrites, characteristics, 41
ferromagnetism, 32–6
  coercive force, 39

ferromagnetism (*cont'd*)
coercivity, 39
hysteresis, 37
hysteresis loop, 39–40
magnetic characteristics, 32, 40
magnetic screening, 37
remanence, 39
remanent flux density, 39
Fleming's rules, 72, 78
left-hand rule, 78
right-hand rule, 72
force exerted on a conductor, 77–8
form factor, 103

generator principle, 73–5

henry, definition of, 81
hysteresis, 37
hysteresis loop, 39, 40

impedance, calculation of, 130
impedance triangle, 129
inductance, mutual, 86
inductance, relationship with physical properties, 84
inductance, self, 79
inductance, unit of, 80
inductive circuit, effect of, 82
inductive reactance, 120
inductive reactance, variation with frequency, 124
inductors, steel-cored, 86
instantaneous value, 101
intrinsic semiconduction, 167

junctions, npn and pnp, 175–6

Kirchhoff's laws, 3, 19

leakage in capacitors, 63
Lenz's law, 69, 75, 77, 79, 143

magnetic characteristics, 32, 40
magnetic fields, 25–7
bar magnet, 26
magnetic flux, 29
magnetic units, 27–31
solenoid, 26

measurements, 140–58
connection of instruments, 145
limitations of instruments, 150
measurement errors, 146, 147
measuring instruments, 141–58
moving coil galvanometer, 141
moving iron galvanometer, 141, 143–5
motor principle, 75
mutual inductance, 86

Newton's second law of motion, 3
null measurement, 151

ohmeter, 156
Ohm's law, 7, 79

peak value of an a.c. wave, 98
periodic time, 101
permeability, 29–34
permittivity, 49–51
phase angle, 109
calculation of, 130
phase difference, 107, 115
phase relationship, 107, 113, 115, 120
phasors, 103
potential difference, 3
potential gradient, 47
potentiometer, 152
power, 4, 123–4
power triangle, 137

ratio of transformer, voltages and turns, 87
rectification, 109–10
reluctance, 31, 34
remanent flux density, 39
resistance, 3, 119
resonance, series, 133

screening, magnetic, 37
semiconductors, 163
biasing, 172
common base connection, 177
common collector connection, 177
common emitter connection, 177

devices, 160
doping, 168–9
intrinsic, 167
junction pn, 169
temperature effect, 164–5
series circuits a.c., 126–8
series circuits d.c., 7–9
sinusoidal waves, 98
algebraic representation, 107
solenoid, magnetic field of, 26
stored energy in capacitance, 58
stored energy in an inductance, 83
superposition theorem, 12, 17

temperature, effect on conductors, 164–5
torque, 4
transformer, basic principles, 4
voltage ratio calculations, 88
voltage and turns ratio, 87
transistor, bipolar, 173
British Standard symbols, 174–5
characteristics, 179
common base, 177
common collector, 177
common emitter, 177
emitter-follower circuit, 178
npn junction, 175
output characteristics, 184
pnp junction, 176
transfer characteristics, 184

units, multiples and sub-multiples, 5

vector quantities, 103
voltage triangle, 129

waveforms, 73, 97
Wheatstones bridge, 151
work, unit of, 4